W9-CSE-791

Characterization
of Solutes in
Nonaqueous Solvents

Symposium on Spectroscopic and.....

Characterization of Solutes in Nonaqueous Solvents

Edited by
Gleb Mamantov
Department of Chemistry
University of Tennessee
Knoxville, Tennessee

Plenum Press • New York and London

CHEMISTRY

6129-6144√

Library of Congress Cataloging in Publication Data

Symposium on Spectroscopic and Electrochemical Characterization of Solute Species in Nonaqueous Solvents, San Francisco, 1976.
Characterization of solutes in nonaqueous solvents.

Includes index.
1. Solution (Chemistry)—Congresses. 2. Spectrum analysis—Congresses. 3. Electrochemical analysis—Congresses. 4. Nonaqueous solvents—Congresses. I. Mamantov, Gleb. II. American Chemical Society. Division of Analytical Chemistry. III. Title.
QD540.S94 1976   541'.3423   77-212204
ISBN 0-306-31108-9

Proceedings of a Symposium on Spectroscopic and Electrochemical Characterization of Solute Species in Nonaqueous Solvents at the American Chemical Society Meeting, Division of Analytical Chemistry, San Francisco, California, August 31—September 1, 1976

© 1978 Plenum Press, New York
A Division of Plenum Publishing Corporation
227 West 17th Street, New York, N.Y. 10011

All rights reserved

No part of this book may be reproduced, stored in a retrieval system, or transmitted, in any form or by any means, electronic, mechanical, photocopying, microfilming, recording, or otherwise, without written permission from the Publisher

Printed in the United States of America

Q D
540
S94
1976
CHEM

Preface

This book consists of contributions by participants in the Symposium "Spectroscopic and Electrochemical Characterization of Solute Species in Non-Aqueous Solvents" which took place at the American Chemical Society Meeting, Division of Analytical Chemistry, August 31 and September 1, 1976, San Francisco, California. The manuscripts were submitted to the editor during the first half of 1977 and, in most cases, represent reviews of selected research topics in the broad area of characterization of solute species in non-aqueous solvents.

In organizing this Symposium, I attempted to bring together a significantly large group of research workers involved in spectroscopic and electrochemical studies in the three large classes of non-aqueous solvents - organic solvents, covalent inorganic solvents and molten salts. The experimental apprcaches and problems, such as avoidance of traces of moisture and oxygen, are frequently similar for all types of non-aqueous solvents.

It is hoped that this volume will be useful to all concerned with chemistry in non-aqueous solvents.

Gleb Mamantov

16324

Contents

IDENTIFICATION AND SYSTEMIZATION OF SOLVENT PROPERTIES INVOLVED IN

THE LIGAND SUBSTITUTION KINETICS OF LABILE COMPLEXES OF NICKEL(II)

J. F. Coetzee, D. Frollini, C. G. Karakatsanis, E. J. Subak, Jr. and K. Umemoto

Department of Chemistry, University of Pittsburgh Pittsburgh, Pennsylvania 15260

1. INTRODUCTION

It is a truism that all reactions occurring in solution are influenced by the solvent. This influence varies from a relatively subtle modulation of solute properties to a profound control over the thermodynamics, mechanisms and kinetics of reactions. Examples of particularly significant solvent effects occur in proton transfer and ligand substitution processes. It is the latter type of process which is the subject of this discussion.

2. LIGAND SUBSTITUTION IN AQUEOUS SOLUTION

The introduction of stopped-flow spectrophotometry and, since about 1950, of relaxation and nuclear magnetic resonance techniques, made it possible to extend kinetic studies from inert to labile complexes (1-8). Until 1967, studies of labile complexes were limited to aqueous solution. The important result emerged that, in aqueous solution, the kinetics of many ligand substitution reactions are relatively insensitive to the nature of the ligand but are similar to the kinetics of solvent exchange in the inner coordination sphere of the metal ion. The results are consistent with an S_N1 or I_d (9) mechanism, i.e., an interchange process with a dissociative mode of activation, in which loss of a solvent molecule from the inner coordination sphere of the metal ion constitutes the rate-determining step. The details of the proposed pathway (1,2) are given in reaction scheme 1:

$$Ni(H_2O)_6^{2+} + L \xrightleftharpoons[k_{21}]{k_{12}} Ni(H_2O)_6^{2+} \cdot L \xrightleftharpoons[k_{32}]{k_{23}} NiL(H_2O)_5^{2+} + H_2O \quad (1)$$

(I) outer-sphere   (II) inner-sphere
complex   complex

in which the rate-determining step is preceded by rapid (diffusion-controlled) formation of an outer-sphere complex. If the ligand is multidentate, additional steps leading to ring-closure must be considered, as illustrated by process III in reaction scheme 2 which applies to an uncharged bidentate ligand:

$$Ni(H_2O)_6^{2+} + L-L \xrightleftharpoons[k_{21}]{k_{12}} (Ni(H_2O)_6^{2+} \cdot L-L \xrightleftharpoons[k_{32}]{k_{23}} NiL-L(H_2O)_5^{2+} + H_2O$$

(I) outer-sphere   (II) singly-coordinated
complex   inner-sphere complex

$$\xrightleftharpoons[k_{43}]{k_{34}} Ni \Big\langle{}^{L}_{L} (H_2O)_4^{2+} + 2H_2O$$

(III) doubly-coordinated (chelated)
inner-sphere complex   (2)

The rate laws for reaction schemes 1 and 2 have been discussed elsewhere (10). Under the simplest conditions (10) and with excess Ni^{2+} present, eq 3 applies:

$$k_{obs} = k_{\ell,f}[Ni^{2+}] + k_{\ell,d} \quad (3)$$

where k_{obs} is the experimentally observed pseudo-first-order rate constant (in units of sec^{-1}), $k_{\ell,f}$ is the second-order rate constant (in ℓ mol^{-1} sec^{-1}) for ligand substitution leading to formation of the inner-sphere complex, and $k_{\ell,d}$ is the first-order rate constant (in sec^{-1}) for dissociation of the inner-sphere complex. Hence, under the simplest conditions, a plot of k_{obs} vs. $[Ni^{2+}]$ is linear with a slope equal to $k_{\ell,f}$. If an I_d mechanism applies to scheme 1 for the reaction of a unidentate ligand, again under the simplest conditions, eq 4 and 5 should be valid:

$$k_{\ell,f} = K_{12}k_{23} \quad (4)$$

and

$$k_{\ell,d} = k_{32} \quad (5)$$

where K_{12} is the equilibrium constant for formation of the outer sphere complex. For the reaction of a bidentate ligand according to scheme 2, the corresponding relationships are

$$k_{\ell,f} = K_{12}k_{23}k_{34}/(k_{32} + k_{34}) \tag{6}$$

and

$$k_{\ell,d} = k_{32}k_{43}/(k_{32} + k_{34}) \tag{7}$$

For the majority of bidentate ligands, ring closure (process III) is sufficiently rapid that $k_{34} \gg k_{32}$, so that eq 6 and 7 reduce to

$$k_{\ell,f} = K_{12}k_{23} \tag{8}$$

and

$$k_{\ell,d} = k_{32}(k_{43}/k_{34}) = k_{32}/K_{34} \tag{9}$$

However, a few bidentate ligands deviate from this "normal" type of substitution. For example, cobalt(II) ion reacts with α-alanine at a normal rate but with β-alanine at a significantly lower rate. This inhibition has been attributed to steric problems in closing the larger ring, and has been termed sterically controlled substitution (11). In such cases, the first bond may form and rupture frequently before ring closure occurs, so that $k_{32} \gg k_{34}$ and eq 6 becomes

$$k_{\ell,f} = K_{12}K_{23}k_{34} \tag{10}$$

Two criteria for a normal I_d mechanism suggest themselves. First, activation parameters for ligand substitution and for solvent exchange should be related as follows:

$$\Delta H^{\ddagger}_{\ell,f} = \Delta H^{\ddagger}_s + \Delta H^0_{12} \tag{11}$$

$$\Delta S^{\ddagger}_{\ell,f} = \Delta S^{\ddagger}_s + \Delta S^0_{12} \tag{12}$$

where ΔH^0_{12} and ΔS^0_{12} are the standard enthalpy and entropy of formation of the outer-sphere complex, and are expected to be small if the outer-sphere complex is relatively unstable, as it usually is. Second, rate constants for ligand substitution and for solvent exchange should be related as follows:

$$k_{\ell,f} = K_{12}k_{23} = K_{12}fk_s \tag{13}$$

where f is a statistical factor representing the probability that the ligand will enter a particular coordination site vacated by a solvent molecule. Choice of a value for f is not a straightforward matter (10); we shall use a value of 3/4. Another limitation of eq 13 is the implicit assumption that solvent exchange is not influenced by the presence of the ligand in the outer sphere. In some cases this is likely to be an oversimplification. It will be convenient to write eq 13 in the form

$$R_1 = (4/3)k_{\ell,f}/K_{12}k_s \tag{14}$$

where the dimensionless ratio R_1 should have a value near unity. It remains to evaluate K_{12}, the equilibrium constant for outer-sphere complexation. If the outer-sphere and inner-sphere complexes are both present in sufficiently high concentrations, both K_{12} and k_{23} in principle will be measurable, e.g., by relaxation methods. However, this condition rarely exists. In most cases, the only possibility is to estimate values of K_{12} from the Eigen-Fuoss equation (12-14):

$$K_{12} = \frac{4\pi Na_1^3}{3 \times 10^3}e^{-U(a_2)/kT} \tag{15a}$$

where

$$U(a_2) = \frac{z_1z_2e^2}{\varepsilon a_2} - \frac{z_1z_2e^2\kappa}{\varepsilon(1 + \kappa a_2)} \tag{15b}$$

Here, a_1 represents the center-to-center distance of closest approach of the solvated metal ion and the reaction site on the ligand, a_2 is the corresponding distance between the two charge sites for the case of a charged ligand and κ is the Debye-Hückel function of the ionic strength, I, given by

$$\kappa^2 = 8 \times 10^{-3}\pi N^2e^2(\varepsilon RT)^{-1}I \tag{15c}$$

where ε is the dielectric constant of the medium. Other symbols have their usual meaning.

It is important to stress the constraints imposed on the application of eq 15. It contains the implicit assumption that no specific interactions occur between the outer and inner spheres. For an uncharged ligand, the electrostatic potential energy, $U(a_2)$, is zero and the remaining term makes no allowance for ion-dipole or dipole-dipole or any other interactions. Consequently, eq 15 is less reliable for uncharged than for charged ligands.

In addition to these inherent uncertainties in the values of K_{12} and f, the application of eq 13 or 14 is further limited by the uncertainty in k_s, the rate constant for solvent exchange. Even greater uncertainties exist in the activation parameters for solvent exchange, as shown in Table I, thereby compromising the application of eq 11 and 12. While there is some basis for preferring certain values in Table I over others, the decisions are not always clear-cut.

As a result of the uncertainties in the application of eq 11, 12 and 14, we propose that, for an individual reaction, no undue significance be attributed to deviations of R_1 from unity by less than a factor of about 10, or to variations in ($\Delta H^{\ddagger}_{\ell,f} - \Delta H^{\ddagger}_S$) by less than 4 or 5 kcal mol^{-1}. Naturally, when trends in kinetic parameters are considered, much smaller variations can be considered

Table I. Rate Constants and Activation Parameters for Solvent

Exchange at Nickel(II) Ion in Various Solvents

Solvent	Method	$\log k_s$	$\Delta H^{\ddagger}_S$ [a]	$\Delta S^{\ddagger}_S$ [b]	Ref.
Water	O-17	4.4	11.6	+ 0.6	15
	O-17	4.5	10.8	- 1.7	16
	O-17	4.5	13.9	+ 8.7	17[c]
	O-17	4.5	12.1	+ 2.9	18
	O-17	4.6	10.3	- 5.2	19
Methanol	H-1	3.0	15.8	+ 8.0	20
Ethanol	H-1	4.0	10.8	- 4	21
Acetonitrile	H-1	3.6	10.9	- 8.8	22
	H-1	3.4	11.7	- 3.6	23
	H-1	4.1	11.8	- 0.2	24
	H-1	3.4	15	+11	25[c]
	N-14	3.3	16.4	+12.0	26
Dimethylsulfoxide	H-1	3.9	8	-14	27
	H-1	4.0	7.3	-16	28
	H-1	3.7	12.1	- 1.3	29[c]
	H-1	3.5	13.0	+ 3.2	30[c]
	H-1	3.9	6.2	-20	31
	H-1	4.0	12.3	+ 1.2	32[c]
Dimethylformamide	H-1	3.6	15	+ 8	33
	O-17	3.9	9.4	- 9.1	34

[a] Units: kcal mol^{-1} (1 cal = 4.18 J).
[b] Units: cal K^{-1} mol^{-1}.
[c] Preferred data; see text.

significant, typically $\pm$ 5% in R_1 and perhaps $\pm$ 2 kcal mol^{-1} in $\Delta H^{\ddagger}_{\ell,f}$, as determined by the experimental accuracy. Much of this may be invalidated by the presence of impurities, especially in relatively inert solvents; this important matter will be discussed in Section 3.3.

Kinetic data for representative ligand substitution reactions in aqueous solution are summarized in Table II. For all but the last two ligands, values of $\Delta H^{\ddagger}_{\ell,f}$ are similar and approach that of $\Delta H^{\ddagger}_s$. Values of R_1 are also similar. The mean value of R_1 is 0.25; it can obviously be brought closer to unity by assuming a value smaller than 3/4 for the statistical factor f in eq 13, but we hesitate to do so in view of the other uncertainties in the calculation of R_1. Deferring consideration of the last two ligands, it is reasonable to conclude that the kinetic data in Table II are consistent with an I_d mechanism. However, it will be shown in subsequent sections that for bipyridine and terpyridine, and possibly also for phenanthroline, this apparently straightforward behavior is the fortuitous result of the protic nature of water.

The last two ligands in Table II are representative of more complex behavior requiring consideration of additional factors, but still within the general framework of an I_d mechanism. For ethylenediamine (and other polyamines), the reaction rate is much higher than the norm. This enhancement in rate has been attributed by Rorabacher (35) to hydrogen bonding of the water molecules of the inner sphere to the ligand molecules present in the outer sphere. Such an interaction, referred to as an <u>internal conjugate base effect</u>, may be expected to increase both K_{12} and k_s, and hence, $k_{\ell,f}$. However, the need to invoke this special mechanism has been questioned because the rate enhancement can be accounted for on the basis of a second pathway involving the monoprotonated ligand (47). Finally, for acetylacetone, the reaction rate is much lower than the norm, and has been attributed to a shift in the rate-determining step from process II to process III of reaction scheme 2 ("sterically controlled substitution").

Finally, an important result emerging from Table II is that the variations of the equilibrium constants, $K_{\ell,f} = k_{\ell,f}/k_{\ell,d}$, of the complexes of some 10 powers of ten are caused almost entirely by variations in $k_{\ell,d}$.

More detailed information about the ligand substitution kinetics of labile complexes in aqueous solution can be found in ref. 1-9.

Table II. Rate Constants and Activation Parameters for Representative Ligand Substitution Reactions of Nickel(II) Ion in Aqueous Solution

Ligand	I	$\log k_{\ell,f}$	$\log R_1$	$\Delta H^\ddagger_{\ell,f}$	$\Delta S^\ddagger_{\ell,f}$	$\log k_{\ell,d}$	$\Delta H^\ddagger_{\ell,d}$	$\Delta S^\ddagger_{\ell,d}$	Ref.
Water	0.3-0.8	4.5[a]		14[a]	+9[a]				17
Ammonia	0.1	3.6	-0.3	10	-6	0.8	13	-10	35,36
Pyridine	0.3	3.6	-0.3	11	-5	1.6	15	+2	36,37
Isoquinoline	<0.03	3.3	-0.6	12	-4	1.5	16	+2	38,39
Bipyridine	<0.015	3.2	-0.7	13	0	-4.3	23	0	37,40,41
Phenanthroline	<0.03	3.5	-0.4	13	+1	-5.0	24	-1	37,40,41
Terpyridine[b]	<0.01	3.1	-0.8	14	+3	-7.6	24	-14	37,40
Thiocyanate[b]	0	4.5	-0.7	12	+3	2.7	18	+13	42
Hydrogen oxalate, HL$^-$	0.1	3.7	-0.9	14	+7	3.2[c]			43
Oxalate, L^{2-}	0.1	4.9	-0.7	14	+12	0.6			43
Ethylenediamine	0.1	5.3	+1.4						44
Acetylacetone (enol form)		0.4	-3.5						45,46

[a] Refers to solvent exchange.
[b] T = 20°C.
[c] In this case, units of $k_{\ell,d}$ are ℓ mol^{-1} sec^{-1}.

3.   EXTENSION TO NONAQUEOUS SOLUTIONS

3.1 Rationale

The solvent may be expected to influence all three steps in
reaction scheme 2. In process II, the assigned role of the solvent
is explicit and crucial. Furthermore, in processes I and III, some
degree of solvent participation is implicit. It therefore may be
hoped that, by proper variation of the solvent, the role of solvent
participation in a range of elementary processes occurring in
solution may be elucidated.

3.2 Systems Studied

During the past few years, a beginning has been made in the
investigation of the solvent dependence of the ligand substitution
kinetics of labile metal complexes. So far, the emphasis has been
on the substitution reactions of nickel(II) ion with uncharged
ligands, since they are sufficiently slow to be studied conveniently
by stopped-flow spectrophotometry, but the reactions of nickel(II)
and several other metals with anionic ligands have also been
studied, mainly by pressure-jump relaxation with conductimetric
monitoring. Typical uncharged ligands studied are shown in Figure
1, and relevant equilibrium constants are listed in Table III.
Solvents have been chosen to allow a wide variation in those properties that could conceivably influence the kinetics of ligand substitution. Such solvents and their properties are listed in Table
IV.

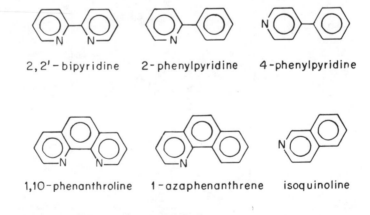

2,2'-bipyridine   2-phenylpyridine   4-phenylpyridine

1,10-phenanthroline   1-azaphenanthrene   isoquinoline

Figure 1.   Typical pyridine-type ligands

Table III. Equilibrium Constants for Formation of Complexes of
Relevant Ligands with Hydrogen Ion and with Nickel(II)
Ion in Various Solvents at 25°C

Ligand	Solvent	$-\log K_a$	$\log K_{\ell,f}{}^a$	Ref.
Ammonia	Water	9.3	2.7,2.2	48
	Acetonitrile	16.5	–	49
Pyridine	Water	5.2	2.1	37
	Acetonitrile	12.3	–	49
	Dimethylsulfoxide	3.4	1.3	40
4-Phenylpyridine	Methanol	–	1.5	50
	Ethanol	–	3.0	50
	1-Propanol	–	3.3	50
	2-Propanol	–	4.0	50
	50-mole% Methanol in 2-Propanol	–	2.4	50
	Dimethylsulfoxide	–	1.5	51
Isoquinoline	Water	–	1.83	39
	Methanol	–	1.70	39
	Dimethylsulfoxide	–	1.49	39
	Dimethylformamide	–	1.96	39
2,2'-Bipyridine	Water	4.3	7.1	37
1,10-Phenanthroline	Water	4.9	8.8	37
2,2',2"-Terpyridine	Water	4.3	10.7	37
Chloride Ion[b]	Dimethylsulfoxide	–	2.7	52
Thiocyanate Ion[b]	Water	–	1.8	42
	Methanol	–	5.2,3.1	42,53
	Acetonitrile	–	5.5,4.3	42
	Dimethylformamide	–	3.9,2.4	42
	Dimethylsulfoxide	–	3,1.5	42,52
Nitrate Ion	Acetonitrile	8.9	4.4[b],3.1[b]	54
Perchlorate Ion	Acetonitrile	–	1.6	55

[a]If only one value is given, it refers to formation of the mono-complex.
If two values are given, the first is that for the mono-complex and the
second is the stepwise formation constant of the bis-complex.
[b]T = 20°C.

Table IV. Selected Properties of Solvents

Property	W	MeOH	EtOH	1-PrOH	2-PrOH	EG	DMF	DMSO	AN	IBN	SL	PC
Freezing temperature, °C	0	-97.8	-114.1	-126.2	-88.0	-12.4	-61	18.45	-45.7	-71.5	28.45	-49
Boiling temperature (1 atm), °C	100	64.5	78.3	97.2	82.3	197.4	153	189	81.6	103.9	287.3[a]	242
Vapor pressure, torr	23.7	120	60	20.85	45	0.14	3.5	0.5	92		5.0[b]	
ΔH of vaporization, kcal mol⁻¹ — at normal bt (1 atm)	9.7	8.4	9.3	10.0	9.5	11.8	10.0	10.7	7.1			
— at 25°C	10.5	9.0	10.1	11.3	10.9	13.8	11.4	12.6	7.7	8.5	15.0[c]	
ΔS of vaporization, cal K⁻¹ mol⁻¹ at normal bt	26.05	24.96				25.1	23.46	23.16	20.06			
Fluidity, g ml⁻¹ cP⁻¹	1.12	1.43	0.73	0.44	0.37	0.061	1.19	0.55	2.27	1.67[d]	0.12[d]	0.47
Relative permittivity	78.5	32.6	24.6	20.3	19.9	37.7	36.7	46.6	36.0	20.4	43.3[d]	64.4
Dipole moment, D	1.85	1.68	1.7	3.09	1.7	2	3.8	4.3	4.1	3.61	4.81	
Refractive index, n_D^{20}	1.333	1.3284	1.3614	1.3856	1.3772	1.4317	1.4306	1.4787	1.344	1.3734	1.4820[d]	1.4209
Donor number	33(?)	19(?)					26.6	29.8	14.1		14.8	14.7

Note: Solvent abbreviations used: Water, W; methanol, MeOH; ethanol, EtOH; 1-propanol, 1-PrOH; 2-propanol, 2-PrOH; ethylene glycol, EG; dimethylformamide, DMF; dimethylsulfoxide, DMSO; acetonitrile, AN; isobutyronitrile, IBN; sulfolane, SL; propylene carbonate, PC. For temperature-dependent properties, temperature is 25°C except where noted otherwise. Data are mainly from Ref. 56 and 57.

[a] Decomposes.
[b] T = 118°C.
[c] T = 100°C.
[d] T = 30°C.

3.3 Solvent Purification and Impurity Effects

The principal experimental problem encountered in the study of fast ligand substitution kinetics is caused by competing ligands (L') present as impurities. The maximum concentration of incoming ligand (L) that can be used in a stopped-flow spectrophotometric experiment is determined by a number of factors, particularly the optical absorptivity of the complex, both the absolute and relative values of the rate constants for the formation and dissociation of the complex, and the minimum permissible concentration of metal ion (eq 3). In the study of mono-complexes it is necessary to avoid the formation of higher complexes by using excess metal ion, and a still greater excess may be necessary if the equilibrium constant for formation of the complex (Table III) is low, or if it is particularly desirable to maintain pseudo-first-order conditions (eq 3). For the ligands listed in Table III, the maximum permissible concentrations under such conditions are typically below 10^{-4} M. Consequently, the level of competing ligands L' present as impurities must be reduced to below 10^{-5} or even 10^{-6} M, depending on the kinetic influence of L'. Typically, if L' is a sufficiently strong ligand as compared to the solvent S to form mixed complexes of the type ML'_aS_b (omitting charges for simplicity), the reaction with L will be accelerated owing to labilization of the remaining solvent molecules in the inner sphere. In such cases, the pseudo-first-order plot of k_{obs} vs. $[M^{n+}]$ (eq 3) may be either nonlinear or, more treacherously, linear but with an intercept on the k_{obs}-axis larger than $k_{\ell,d}$. Naturally, such problems are most severe in solvents of low donor strength, e.g., propylene carbonate and sulfolane.

Purification procedures used for the solvents of interest here are listed in Table V. Information on residual impurities is sketchy; what is known is also included. We have discussed the kinetic effects of particular impurities elsewhere (10, 62, 65).

3.4 Results

The first systematic studies of labile complexes in nonaqueous solvents were carried out by Pearson and Ellgen (66) in 1967 (nickel(II) with uncharged and anionic ligands in methanol) and by Bennetto and Caldin(56) in 1971 (mainly nickel(II) with bipyridine in seven solvents). Further work by several authors with a variety of ligands and solvents followed. We have reviewed the status of the field through 1974 (10). In the present discussion, we shall emphasize those systems characterized by "abnormal" kinetics. Results obtained in dimethylsulfoxide and acetonitrile are of particular interest and are shown in Tables VI and VII. Complementary information is given in Table VIII, where rate constants of pyridine-type ligands in all solvents studied so far are analyzed.

Table V. Purification Procedures Used for Solvents, and Known Residual Impurities

Solvent	Principal Impurities in Technical Grade Solvent[a]	Purification Procedure and Remaining Impurities	Ref.
EG	1,2-Propanediol, butanediol, diglyme, ethylene oxide, lower alcohols.	Distill in vacuo from anhydrous $CaSO_4$. Water, 15-20 mM.	56
DMSO	Dimethylsulfide, dimethylsulfone (trace), unknown organic impurity.	Distill in vacuo, first alone, then from CaH_2. Water, ca. 4 mM.	40,58
DMMP		Dry over molecular sieves, then distill twice in vacuo from anhydrous $CaSO_4$.	56
DMF	Formic acid, dimethylamine, hydrogen cyanide, carbon dioxide, carbon monoxide.	Dry over molecular sieves, then treat alternately with KOH and P_2O_5 and distill three times in vacuo from P_2O_5.	38,59
MeOH	Carbon dioxide, methyl ether, methylal, methylol, methyl formate, methyl acetate, formaldehyde, acetaldehyde, acetone and various lower alcohols.	Distill from Mg turnings under nitrogen. Water, ca. 8 mM.	
EtOH	Aldehydes, ketones and other oxygenated compounds.	As for MeOH. Water, ca. 5 mM.	
1-PrOH	2-Propene-1-ol	As for MeOH. Water, ca. 20 mM.	
2-PrOH		As for MeOH. Water, ca. 30 mM.	
AN	Acetic acid, ammonia, unsaturated nitriles (acrylonitrile).	Dry with silica gel, distill first from P_2O_5 and then from CaH_2. Water, ca. 1 mM; acrylonitrile, < 1 mM.	60
IBN	Methacrylonitrile, acidic impurities (such as isobutyric acid formed by hydrolysis of the solvent).	Shake successively with silica gel and neutral alumina, then distill, first from P_2O_5 and then from CaH_2. Water, ca. 8 mM.	64,60
SL	3-Sulfolene (dissociates into sulfur dioxide and 1,3-butadiene at temperatures above 70°C); 2-sulfolene and isopropyl sulfolanyl ether may also be present.	Distill in vacuo, twice from NaOH and then by itself. Water, 3-5 mM.	61
PC	Propylene oxide, allyl alcohol, 1,2-propanediol (propylene glycol), 1,3-propanediol, carbon dioxide, propionaldehyde.	Dry over molecular sieves, then distill in vacuo, twice from anhydrous $CaSO_4$ and once by itself. Water, ca. 4 mM; propylene glycol, < 0.03 mM.	62,63

[a] Water is not listed, but it is understood that it is always present.

Table VI. Comparison of Rate Constants and Activation Parameters

for Ligand Substitution Reactions in Dimethylsulfoxide

with Corresponding Quantities for Solvent Exchange

Ligand	$\log k_s$ or $\log k_{\ell,f}$	$\log R_1$	$\Delta H^{\ddagger}_s$ or $\Delta H^{\ddagger}_{\ell,f}$	$\Delta S^{\ddagger}_s$ or $\Delta S^{\ddagger}_{\ell,f}$	Ref.
Dimethylsulfoxide	3.7[a]		12[a]	+1[a]	
Chloride ion	4.8[b]	-0.1			52
Thiocyanate ion	4.9[b]	0.0			52
Pyridine	3.4	-0.1			40
Isoquinoline	3.5	0.0	8	-17	38
4-Phenylpyridine	3.2	-0.3	9	-13	51
Bipyridine	1.84	-1.7	13	-8	56,67
Phenanthroline	2.6	-0.9	11	-8	10
Terpyridine	1.41	-2.1	12	-12	56,67
	1.40	-2.1	15	-2	40
2-(Aminomethyl)pyridine	3.64	+0.1	9	-11	68
2-(2-Aminoethyl)pyridine	3.65	+0.1	12	0	68
2-(Aminomethyl)-6-methylpyridine	3.73	+0.2	14	+7	68
2-[(Methylamino)methyl]pyridine	3.33	-0.2	9	-14	68
PADA[c]	2.87	-0.6	12	-4	69

[a]Somewhat uncertain; see text and Table I.
[b]$T = 20°C$.
[c]Pyridine-2-azo-p-dimethylaniline.

Table VII. Comparison of Rate Constants and Activation Parameters

for Ligand Substitution Reactions in Acetonitrile with

Corresponding Quantities for Solvent Exchange

Metal Ion	Ligand	$\log k_s$ or $\log k_{\ell,f}$	$\log R_1$	$\Delta H^{\ddagger}_s$ or $\Delta H^{\ddagger}_{\ell,f}$	$\Delta S^{\ddagger}_s$ or $\Delta S^{\ddagger}_{\ell,f}$
Nickel(II)	Acetonitrile	3.43	-	15	+11
	Thiocyanate ion[a]	5.00	-0.1	17[b]	+22[b]
	Nitrate ion[a]	5.28	+0.1	17.5[b]	-
	Trifluoroacetate ion[a]	5.18	0	-	-
	p-Toluenesulfonate ion[a]	5.11	0	17.0[b]	-
	Ammonia	3.5	+0.2	-	-
	Pyridine	2.92	-0.4	14.7	+4
	4-Phenylpyridine	2.99	-0.3	11.2	-7
	Isoquinoline	3.09	-0.2	11.9	-5
	2,2'-Bipyridine	3.61	+0.3	6.5	-20
	2,2',2"-Terpyridine	3.34	0.0	8.4	-15
	1,10-Phenanthroline	4.70	+1.4	4.7	-21
	5-Nitrophenanthroline	4.09	+0.8	5.1	-23
	5-Chlorophenanthroline	4.37	+1.1	6.5	-17
	5-Methylphenanthroline	4.81	+1.5	5.5	-18
	5,6-Dimethylphenanthroline	4.86	+1.6	3.5	-25
	2,9-Dimethylphenanthroline	2.63	-0.7	10.1	-13
	PADA[c]	4.23	+0.9	5.9	-19

Note: Data are from Ref. 70 except where indicated otherwise. Uncertainty in $\Delta H^{\ddagger}$ is estimated to be ±1 kcal mol^{-1}; however, see Section 3.3.

[a]Data are from Ref. 42 and 54 for T = 20°C.
[b]Comparison of the values for anionic ligands with those for uncharged ligands requires correction for the ΔH^0 and ΔS^0 values for the formation of the outer-sphere complex (eq. 11, 12), viz., 2.2 kcal mol^{-1} and 15.5 cal K^{-1} mol^{-1}, respectively (Ref. 42). This correction leads to the following mean values for the rate-determining step for the three anionic ligands: $\Delta H^{\ddagger}_{23} = 15.0 \pm 0.5$, $\Delta S^{\ddagger}_{23} = 6.5 \pm 2$.
[c]Pyridine-2-azo-p-dimethylaniline; data are from Ref. 69.

Table VIII. Comparison of Rate Constants for Substitution by Phenanthroline, Bipyridine and Terpyridine at Nickel(II) Ion with Rate Constants for Solvent Exchange and for Substitution by Isoquinoline or 4-Phenylpyridine

Solvent	D^a	$\log k_s^b$	Thiocyanate[c]	Iso-quinoline[d]	4-Phenyl-pyridine[d]	Phenanthroline[d]	Bipyridine[d]	Terpyridine[d]	Ref.
EG		3.6?	2.7, 0.5	1.3, -2.0	1.2, -2.0	2.2, -1.0, +1.0	1.5, -1.8, +0.3	0.8, -2.4, -.04	71
DMSO	30	3.7	4.9[e], 3.4	3.5, 0.0	3.2, -0.3	2.6, -0.9, -0.6	1.8, -1.7, -1.4	1.4, -2.1, -1.8	T VI
DMMP		4.4					2.3, -2.0,		56
DMF	27	3.9					2.7, -1.1, -0.7[g]		38,56
W	33?	4.5	5.1[e], 3.2	3.4, -0.3	3.6[f], -0.3	3.5, -0.4, -0.1	3.2, -0.7, -0.4	3.1, -0.8, -0.5	T II
MeOH	19	3.0	4.5[e], 3.2	3.3, -0.6	2.1[f], -0.5	2.8, +0.2, +0.7	2.0, -0.6, -0.1	1.6, -1.0, -0.5	72
EtOH		4.0	5.0[e], 2.5	2.0, -0.6	3.8[f], -0.1	4.5, +0.6, +0.7	3.8, -0.1, 0		73,74
1-PrOH					4.2,	5.1, , +0.9	4.0, , -0.2		75
2-PrOH					5.2,	5.9, , +0.7	, , -0.2	4.4, , -0.8	75
AN	14	3.4	5.0[e], 3.2	3.1, -0.2	3.0, -0.3	4.7, +1.4, +1.7	3.6, +0.3, +0.6	3.3, 0.0, +0.3	T VII
IBN					3.9,	>5, , >1	3.7, , -0.2	3.6, , -0.3	75
SL	15				4.5,	ca.6, , ca.1.5	4.6, , +0.1	4.3, , -0.2	76
PC	15				5.3,	7, , +1.7	5.4, , +0.1	5.3, , 0	77

Note: Solvent abbreviations used: Ethylene glycol, EG; dimethylsulfoxide, DMSO; dimethylmethylphosphonate, DMMP; dimethylformamide, DMF; water, W; methanol, MeOH; ethanol, EtOH; 1-propanol, 1-PrOH; 2-propanol, 2-PrOH; acetonitrile, AN; isobutyronitrile, IBN; sulfolane, SL; propylene carbonate, PC. T = 25°C except where noted otherwise.

[a] Donor number.
[b] Refers to solvent exchange.
[c] First entry represents $\log k_{\ell,f}$; second entry represents $\log (k_{\ell,f} \times \frac{K_{12,4\text{-phenpyr}}}{K_{12,SCN^-}})$. K_{12} was calculated from the Eigen-Fuoss equation with the following values of a (in Å): EG; DMSO, 7; DMF, 7; W, 5; MeOH, 6; and AN, 7.5, all at I = 0.
[d] For each ligand-solvent combination, the first entry represents $\log k_{\ell,f}$, the second $\log R_1$, and the third $\log R_2$ where

$$R_1 = (4/3)k_{\ell,f}/K_{12}k_s \text{ and } R_2 = (k_{\ell,f})_L/(k_{\ell,f})_{L'},$$ with L' representing 4-phenylpyridine as reference ligand. K_{12} was calculated as in c above, with the following additional values of a: DMMP, 7.5; EtOH, 7.
[e] T = 20°C.
[f] Ligand is pyridine.
[g] In this case, reference ligand is isoquinoline.

Sources of data for water, dimethylsulfoxide and acetonitrile are given in Tables II, VI and VII. Solvents are listed in the order that we believe to be that of decreasing donor ability. We expect this order to deviate from that of the donor numbers for reasons explained elsewhere (78).

3.5   Towards Understanding the Role of the Solvent

In Section 1, criteria for a dissociative interchange mechanism were established (eq 11, 12, 14). These criteria are now applied in Tables VI and VII. Uncertainties in this diagnostic approach were also discussed in Section 1. Consequently, we shall refrain from attributing much significance to small deviations of $\Delta H^{\ddagger}_{\ell, f}$ from $\Delta H^{\ddagger}_{s}$, or of R_1 from unity. We shall rather focus our attention on <u>trends</u> in rate constants and activation parameters. This is done more formally in Table VIII, which also introduces an additional dimensionless parameter, R_2, given by

$$R_2 = (k_{\ell, f})_L / (k_{\ell, f})_{L'} \tag{16}$$

where L' is 4-phenylpyridine, chosen as reference ligand. This choice is justified, because Table VIII shows that the R_1-values of 4-phenylpyridine are "normal" in all solvents for which exchange rates are available, except ethylene glycol; but for this bidentate ligand, it is not clear what k_s represents. This additional criterion for "normal" substitution, <u>viz</u>., $R_2 \sim 1$, is the only one available for those systems for which solvent exchange parameters are not yet available.

The main features of the data in Tables VI - VIII are the following.

1.   The apparently straightforward behavior of aqueous solutions is not general. In particular, the kinetic properties of ligands are much more differentiated in nonaqueous solvents than in water. While unidentate ligands show only a modest degree of specificity and conform to the criteria for a simple I_d mechanism, as they do in water, multidentate pyridines show much greater specificity characterized by either positive or negative deviation from the norm in several solvents.

2.   In all solvents except water and dimethylsulfoxide, phenanthroline reacts faster than the norm. In all solvents, without exception, phenanthroline reacts faster than bipyridine does; this enhanced reactivity of phenanthroline is small in water but much greater in all nonaqueous solvents, particularly in acetonitrile, sulfolane and propylene carbonate.

3.   In dimethylsulfoxide, phenanthroline and particularly bipyridine and terpyridine react much more slowly than the norm. Furthermore, while the other 2-substituted pyridines listed in Table VI react at normal rates at 25°C, their activation parameters reflect a substantial degree of ligand specificity.

We shall consider first the abnormally slow reactions of phenanthroline, bipyridine and terpyridine in dimethylsulfoxide as solvent. The possibility of a limiting dissociative (lim S_N1 or D) mechanism characterized by a relatively long-lived 5-coordinate intermediate must be considered, but it seems unlikely because the unidentate ligands do not show the specificity expected for a D mechanism.

It will be shown that the orientation of these ligands in the outer sphere is of crucial importance. In the solid state, the rings of bipyridine are coplanar but in the trans configuration (79). Even in solution (benzene as solvent), the average orientation of the rings approaches the trans configuration, as shown by the low dipole moment of the molecule, given in Table IX. Calculations show that the completely cis and completely trans configurations should have moments of 3.8 and ca. 0 D, respectively, while free rotation would result in a moment of 2.7 D (83). The observed moment of the rigid molecule phenanthroline, in which the nitrogen atoms are fixed in the cis configuration, is in close agreement with the value calculated for the cis configuration of bipyridine. For terpyridine, the preferred configuration is trans-trans.

Table IX. Dipole Moments of Ligands

Ligand	μ, D[a]	Ref.
Pyridine	2.37[b]	80
Isoquinoline	2.73[c]	81
2-Phenylpyridine	1.77	82
3-Phenylpyridine	2.45	82
4-Phenylpyridine	2.50	82
2,2'-Bipyridine	0.69	80
1,10-Phenanthroline	3.64	80

[a]Measured in benzene, except where noted otherwise.
[b]Measured in carbon tetrachloride.
[c]Measured in gas phase.

Two hypotheses have been proposed to account for the slowness of the reactions of bipyridine and terpyridine in dimethylsulfoxide as solvent. Coetzee et al. (40, 84) invoked a shift in the rate-determining step from first-bond formation to ring closure (process III of reaction scheme 2). It was suggested that the strongly developed sheet-like structure of liquid dimethylsulfoxide (10) may present a barrier to rotation of the free ring(s) of the singly-coordinated intermediate from the trans to the cis configuration. On the other hand, Moore showed subsequently (68) that the other 2-substituted pyridines listed in Table VI, and which presumably also act as bidentate ligands, react much faster than bipyridine does. Moore accounted for these observations by invoking steric hindrance during first-bond formation by bipyridine and terpyridine as a result of interference between the adjacent (3') hydrogen atom on the flanking trans pyridine ring and the bulky dimethylsulfoxide molecules present in the inner sphere.

We have been able to show subsequently (78) that neither ex-planation is quite correct, although Moore's is nearer the truth. The key to the solution of the problem is the behavior of the ligand 2-phenylpyridine. The phenyl ring causes such severe steric problems that this compound shows virtually no tendency to complex with nickel(II) ion in water, in dimethylsulfoxide, or even in the weaker donors 2-propanol and acetonitrile. However, it does react in the very weak donors propylene carbonate and sulfolane; details will be presented elsewhere. In propylene carbonate it actually forms a moderately stable complex, the equilibrium constant (determined spectrophotometrically) being ca. 1×10^3 ℓ mol^{-1}. How-ever, the rate of formation of the complex is exceptionally low with $k_{\ell,f} \sim 2$ ℓ mol^{-1} sec^{-1}, or some 5 powers of ten lower than the values for 4-phenylpyridine and bipyridine (Table VIII). Hence, even in propylene carbonate, formation of the 2-phenylpyridine complex is subject to severe steric hindrance, to a degree that is not en-countered by the bipyridine complex. Parenthetically, these steric problems do not cause unusually fast dissociation of the complex, $k_{\ell,d}$ being of the order of 10^{-3} sec^{-1}.

Dimethylsulfoxide is a much stronger donor than propylene carbonate towards nickel(II) ion, so that, when increasing amounts of dimethylsulfoxide are added to a solution of nickel(II) ion in propylene carbonate, the reaction

$$\text{Ni(PC)}_6^{2+} + n\text{DMSO} \rightleftharpoons \text{Ni(DMSO)}_n\text{(PC)}_{6-n}^{2+} + n\text{PC} \qquad (17)$$

proceeds almost stoichiometrically up to n = 5. Now, the effect of DMSO in the inner sphere of nickel(II) ion is much more drastic for the reaction of 2-phenylpyridine than for that of bipyridine, as shown in Figure 2. Also included are corresponding data for 4-phenylpyridine, which behaves normally in pure dimethylsulfoxide

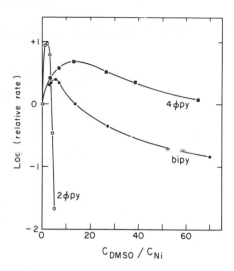

Figure 2.   Effect of added dimethylsulfoxide on rates of reaction of nickel(II) ion with 2-phenylpyridine, 4-phenylpyridine and
2,2'-bipyridine in propylene carbonate as solvent.   For 2-phenyl-
pyridine, $C_L = 5 \times 10^{-5}$, $C_{Ni} = 10 \times 10^{-3}$ M; for the other two
ligands, $C_L = 2.5 \times 10^{-5}$, $C_{Ni} = 1 \times 10^{-3}$ M.   Initial rates have
been normalized; see text.   From ref. 78.

and therefore may serve as a reference ligand.   The reaction rate
of 4-phenylpyridine passes through a maximum at $C_{DMSO}/C_{Ni} \equiv R_c \sim 10$.
There is little doubt that, at this solvent composition, n = 5
(eq 17), because labilization of propylene carbonate by the stronger
donor dimethylsulfoxide will reach a maximum at this value of n
(48,77).   Parenthetically, it also appears that introduction of a
sixth molecule of dimethylsulfoxide occurs with some difficulty,
probably because steric crowding becomes significant.

For bipyridine, the reaction rate reaches a maximum at $R_c \sim 5$
or 6, when (probably) n = 4.   Now, it is particularly significant
that for 2-phenylpyridine the maximum rate occurs at a much lower
value of R_c, between 1 and 2, and that the rate has dropped precipi-
tously before R_c (hence, the maximum possible value of n) has
reached a value of 3 or 4.   We conclude that inhibition of the reactivity of 2-phenylpyridine by addition of dimethylsulfoxide is the
result of exhaustion of the acceptor ability of nickel(II) ion by
the high donor strength of dimethylsulfoxide, rather than of steric
inhibition caused by the bulkiness of the dimethylsulfoxide molecules, as suggested by Moore (68).   This conclusion is supported by
the fact that addition of acetonitrile, which actually alleviates
steric crowding in the inner sphere, produces similar trends in the
reaction rate, but occurring at higher concentration of cosolvent,
exactly as would be expected because acetonitrile is a weaker donor

than dimethylsulfoxide. When R_C reaches a value of 50 for acetonitrile, or 10 for dimethylsulfoxide, there is no longer any
observable reaction.

When further dimethylsulfoxide is added in concentrations
higher than those shown in Figure 2, the reaction rates of both
4-phenylpyridine and particularly bipyridine continue to decrease.
Since there is no concomitant decrease in solvent exchange rate
(85), we have attributed this decrease to gradual exclusion of the
incoming ligand from the outer sphere by dimethylsulfoxide (78).

The relevance of these results to the abnormally slow reactions
of bipyridine and terpyridine in dimethylsulfoxide is that the
kinetic properties of 2-phenylpyridine should simulate those of
2,2'-bipyridine during first-bond formation for all configurations
of bipyridine except cis. We conclude that bipyridine will form
its first bond only when its two rings are in the cis configuration,
because bonding will be subject to severe steric hindrance when the
second ring is substantially out of the cis position. Second bond
formation should be fast. For terpyridine, similar considerations
should apply to formation of the first and second bonds, but formation of the third bond could be slower and become rate limiting.
For phenanthroline, the two donor atoms are already held rigidly in
the cis configuration, so that its reactivity should be higher than
that of bipyridine, as observed.

These considerations rule out our earlier suggestion (40, 84)
that for bipyridine the rate-determining step is ring closure.
Furthermore, we have subsequently found no evidence for the fast
formation of a singly-coordinated intermediate, which should absorb
at lower wavelengths than the chelate does. We therefore agree
with Moore (68) that first-bond formation is rate determining and
that it is inhibited sterically, but unlike Moore, we attribute the
steric inhibition to the need for proper orientation of the two
rings of the ligand in the outer sphere.

The above conclusions came from a consideration of certain
abnormally slow reactions in dimethylsulfoxide. They will be extended to other solvents after certain abnormally fast reactions
have been considered. The most thoroughly studied examples of this
type are bipyridine, terpyridine and especially the phenanthrolines
in acetonitrile as solvent, as shown in Table VII. Not only are
values of R_1 greater than unity, but values of $\Delta H^{\ddagger}_{\ell,f}$ are much
smaller than those of $\Delta H^{\ddagger}_s$. We have attributed this enhanced
reactivity to exceptional stabilization of the outer-sphere complexes (40, 70). The possibility of an associative mechanism must
be considered, but it can be ruled out because the sterically less
demanding unidentate ligands show no enhanced reactivity. Also,
there is no spectral or other evidence for any deviation of the
solvated nickel(II) ion from octahedral symmetry.

A variety of interactions may contribute to the proposed
stabilization of the outer-sphere complexes. We believe that in
aprotic solvents the principal interaction is of the ion-dipole
type, occurring between the effective positive charge of the metal
ion and the dipole of the ligand. Such interactions will of course
be strongest for ligands having the highest dipole moments, such as
phenanthroline (see Table IX) and it will be strongest in those
solvents having the weakest donor ability (so that the effective
positive charge of the metal will be highest) and in which the
inner sphere of the metal ion is thinnest and/or most open (so that
the interaction distance can be small). The acetonitrile molecule
has only modest steric requirements laterally, so that the inner
sphere of nickel(II) ion in that solvent is sufficiently open to
allow considerable penetration by the ligand. In propylene carbonate and particularly in sulfolane, the inner sphere of nickel(II)
ion is less open than in acetonitrile, but the donor abilities of
these solvents are lower than that of acetonitrile, so that the net
result is similar enhancement of the reactivity of phenanthroline
(Table VIII).

It is likely that other types of interaction also contribute
to the stabilities of outer-sphere complexes in aprotic solvents,
but insufficient information is available to assess their significance. For example, short-range interactions of the dipole-dipole
and dispersion types between the ligand and the solvent molecules
of the inner sphere may be significant, particularly because the
normal polarity of the solvent molecules will be increased by the
field of the ion.

In protic solvents, hydrogen bonding of the polarized solvent
molecules of the inner sphere to the ligand will be particularly
important.

Whatever the details may be of the interactions occurring between the outer and inner spheres, there can be little doubt that
the outer-sphere complexes of the phenanthrolines in particular
possess "extra" stability in acetonitrile. In addition to the
abnormally high rate constants and low enthalpies of activation for
the reactions of these ligands, two other lines of evidence support
this hypothesis. First, the rate constants of 5- and 6-substituted
phenanthrolines obey the Hammett correlation shown in Figure 3.
Since the substituents are remote from the donor atoms, their
effect cannot be steric but can only be electronic in nature.
Electron-donating substituents increase the electron density at the
donor atoms and therefore increase the interactions stabilizing the
outer-sphere complex, and vice versa (70). The second, and more
compelling, line of evidence comes from the kinetic effect of
anionic substituents (X) in the inner sphere of nickel(II). In
Table X, rate constants for the two overall reactions

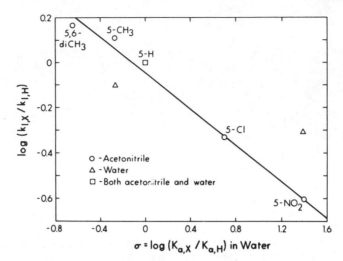

Figure 3.   Hammett correlation for reactions of substituted
phenanthrolines with nickel(II) ion in acetonitrile as solvent.
Slope = -0.40 ± 0.03.   From ref. 70.

$$NiS_6^{2+} + Y \rightleftharpoons NiYS_a^{2+} \qquad (k_{Ni}) \qquad (18)$$

and

$$NiXS_b^{+} + Y \rightleftharpoons NiXYS_c^{+} \qquad (k_{NiX}) \qquad (19)$$

are compared for various incoming ligands (Y) in the two nearly
isodielectric solvents S = acetonitrile (65) and S = methanol (72).
In all cases, the presence of X in the inner sphere accelerates
reaction with Y, i.e., $k_{NiX}/k_{Ni} \equiv N > 1$.   This acceleration is the
net result of two opposing effects of X: labilization of the
solvent molecules in $NiXS_b^{+}$, and destabilization of the outer-sphere
complexes $NiXS_c^{+}Y$.   The latter factor should be most important for
that ligand forming the most stable outer-sphere complex, $NiS_6^{2+}Y$,
i.e., for Y = phenanthroline.   This expectation is fulfilled: the
acceleration number, N, is smallest for phenanthroline in both
solvents (compare entries 5, 9 and 10).   Parenthetically, the
relative acceleration numbers for the two solvents, N_{AN}/N_M, parallel
the free energies of transfer of the free anions from methanol to
acetonitrile.   This is the expected behavior.   All four anions are
more weakly solvated, and therefore have higher activities, in
acetonitrile than in methanol.   This increased activity in the
aprotic solvent is largest for the "hard" anion, chloride ion,
which prefers hydrogen bonding for stabilization, and it is smallest
for the "soft" anion, thiocyanate ion, which is more capable of

Table X. Accelerating Effect of Anionic Inner-Sphere Substituents $(\underline{X})$ on Rates of Reactions of Nickel(II) with Various Incoming Ligands $(\underline{Y})$ in Methanol and Acetonitrile.

Y	X	Methanol			Acetonitrile				
		$\log k_{Ni}$	$\log k_{NiX}$	$\log N^{\underline{a}}$	$\log k_{Ni}$	$\log k_{NiX}$	$\log N^{\underline{a}}$	$\log(N_{AN}/N_M)$	$\Delta G_t^\circ(X^-)^{\underline{b}}$
4-Phenylpyridine	1. Cl	2.1	3.9	1.8					
	2. Br		3.3	1.2					
	3. I		2.8	0.7					
	4. SCN		2.9	0.8					
Bipyridine	5. Cl	2.0	3.1	1.1	3.61	6.3	2.7	1.6	5.7
	6. Br		2.5	0.5					
	7. I		2.3	0.3					
	8. SCN		3.0	1.0					
Phenanthroline	9. Cl	2.8	3.1	0.3	4.70	6.6	1.9	1.6	5.7
Terpyridine	10. Cl	1.6	2.8	1.2	3.38	6.3	2.9	1.7	5.7
	11. Br		2.2	0.6		5.7	2.3	1.7	4.9
	12. I		1.8	0.2		4.8	1.4	1.2	2.9
	13. SCN		2.8	1.2		5.5	2.1	0.9	1.6
	14. NO₃					4.4	1.0		

$^{\underline{a}}N = k_{NiX}/k_{Ni}$. Data are from Ref. 65 and 72.
$^{\underline{b}}$Free energy of transfer of free X^- from methanol to acetonitrile, in kcal mol^{-1}; from Ref. 86.

stabilization by dispersion interactions. Hence, it is to be expected that the Ni-X bond will be stronger and therefore the solvent molecules in $NiXS_b^+$ will be labilized more in acetonitrile than
in methanol, as observed.

In summary, we attribute both the enhanced reactivity of
certain ligands in acetonitrile and the inhibited reactivity of
certain ligands in dimethylsulfoxide to variations in outer-sphere
stabilities and configurations determined by an interplay of the
steric requirements and effective polarity of the ligand, and the
steric requirements and donor ability of the solvent.

Another theory for solvent participation in ligand substitution reactions has been proposed before. Bennetto and Caldin (56,
67) observed correlations between the reactivity of bipyridine and
certain properties of the solvent, such as its fluidity and enthalpy of vaporization, which can be related to its structure.
These correlations served as the basis of a theory for ligand substitution in which the structure of the solvent was proposed to be
the crucial factor. It now seems unnecessary to attribute that
much significance to solvent structure as such. We have shown here
that bipyridine has severe and atypical steric requirements.
Furthermore, those solvents included in Bennetto and Caldin's correlations having low fluidities and high enthalpies of vaporization
(dimethylsulfoxide, dimethylmethylphosphonate, and dimethylformamide) also are strong donors and have bulky molecules, and it is
these latter properties that are of primary significance. At the
other end of the spectrum, while acetonitrile does have a high
fluidity and low enthalpy of vaporization, its kinetically significant properties are its weak donor ability and modest steric
requirements. This de-emphasis of the kinetic role of solvent
structure per se is supported by the observation by Langford (87)
that solvent exchange is insensitive to bulk solvent structure when
the entering group, leaving group and nonlabile ligands are all
kept invariant.

4.   CONCLUSION

A Working Hypothesis

The ligand substitution kinetics of the labile complexes of
nickel(II) and several other metals (10) in all solvents studied
so far can be accommodated within the framework of the dissociative
interchange (I_d) mechanism illustrated in reaction schemes 1 and 2.
The salient kinetic feature of aqueous solutions is an apparent
lack of ligand specificity, but, as for proton transfer and other
classes of chemical reactions, the properties of aqueous solutions
are atypical. In nonaqueous solvents, ligand specificity is more

pronounced.   Ligands such as phenanthroline, bipyridine and terpyridine show particularly specific behavior.   In certain solvents,
reactivity of these ligands is enhanced above the norm of substitution by "simple" unidentate ligands, while in other solvents
their reactivity is inhibited.

We propose the following working hypothesis to systematize the
substitution kinetics of nickel(II) ion in different solvents.

1.   The norm is an unperturbed I_d mechanism for which eq 11-15
apply.
2.   Deviations from the norm originate mainly in the outer-
sphere complexation step.
3.   Both the stability of the outer-sphere complex (represented by K_{12} in eq 13-15) and the orientation of the ligand in
the outer sphere are important.
4.   Reactivity of a pyridine-type ligand is influenced by its
dipole moment and steric requirements.   Bipyridine exists in solution with its two rings preferentially in the trans configuration,
but its steric requirements are such that it can form even its
first bond only when its rings are oriented in the cis configuration.   Solvent properties favoring such proper orientation are
those that would promote interaction of the ligand with the inner
sphere (e.g., hydrogen bonding ability) or with the "effective"
positive charge of the metal ion (low donor strength and small
size).
5.   The following operational classification of solvents is
useful in systematizing the deviations from the norm shown by the
data in Table VIII.

   I.   The Solvent is a Weak Donor

      A.   The Inner Sphere is Relatively Open

         Examples: acetonitrile, sulfolane, propylene car-
         bonate.   Ion-dipole interaction of the metal ion with
         the ligand in the outer sphere is at a maximum.   The
         reactivity of phenanthroline (and even of bipyridine
         and terpyridine in acetonitrile) is enhanced to a
         maximum extent.

      B.   The Inner Sphere is Relatively Closed

         Example: isobutyronitrile.   The effects characteriz-
         ing class I.A are diminished.

  II.   The Solvent is a Strong Donor

      A.   The Solvent is Aprotic

Examples: dimethylsulfoxide, dimethylmethylphosphonate, dimethylformamide. In these solvents, particularly the first two, the fact that the inner spheres are relatively closed also contributes to inhibition of reaction rates. Reactivity of bipyridine and terpyridine (and even of phenanthroline in dimethylsulfoxide) is inhibited to the maximum extent.

B.   The Solvent is Protic

Examples: ethylene glycol, water, methanol, ethanol, 1-propanol, 2-propanol. The inhibiting effect of the relatively high donor strength of the solvent is offset by the enhancing effect of hydrogen bonding of the inner sphere to the ligand. A special case is that the reaction rate of phenanthroline in aqueous solution is leveled; a possible reason may be the well-established specific interaction of phenanthroline with water, which actually results in the formation of a stable monohydrate (40, 88).

5.   SUMMARY

The ligand substitution kinetics of nickel(II) ion have been studied in a variety of solvents of widely different properties. All results can be accommodated within the framework of a broadly-defined dissociative interchange mechanism in which solvent exchange contributes to the kinetics. However, the apparently straightforward behavior of aqueous solutions, in which most ligands show little specificity, is not general. In certain nonaqueous solvents, the reactivities of such ligands as phenanthroline, bipyridine and terpyridine are enhanced, while in other solvents they are inhibited, relative to norms established for "simple" reactions. It is shown that such deviations from the norm originate in the outer-sphere complexation step of the reaction pathway. Not only the stability of the outer-sphere complex, but also the orientation of the ligand in the outer sphere is important. For example, it is shown that, probably in all solvents, the flexible ligand bipyridine can form even its first bond only when its rings are in the cis configuration. Solvent properties responsible for deviations from the kinetic norm include donor strength and molecular size. An operational classification of solvents is presented to systematize solvent effects in substitution kinetics.

6.   ACKNOWLEDGMENTS

All work done in our laboratory was supported by the National Science Foundation. We also acknowledge information provided to us before publication by Professors B. Kratochvil and W. M. Smith.

The data manipulation necessary to obtain rate constants and activation parameters from our recent stopped-flow experiments was performed using appropriate computer programs giving least-squares fits. These were written by one of us (EJS).

7. REFERENCES

(1) M. Eigen, Z. Elektrochem., 64, 115 (1960); Pure and Applied Chem., 6, 97 (1963).

(2) M. Eigen and R. G. Wilkins, Advan. Chem. Ser., No. 49, 55 (1965).

(3) F. Basolo and R. G. Pearson, "Mechanisms of Inorganic Reactions," Wiley, New York, N.Y., 1967.

(4) A. McAuley and J. Hill, Quart. Rev. Chem. Soc., 23, 18 (1969).

(5) D. J. Hewkin and R. H. Prince, Coord. Chem. Rev., 5, 45 (1970).

(6) R. G. Wilkins, Accounts Chem. Res., 3, 408 (1970).

(7) C. H. Langford, in S. Petrucci, ed., "Ionic Interactions," Academic Press, New York, N.Y., 1971.

(8) R. G. Wilkins, "The Study of Kinetics and Mechanism of Reactions of Transition Metal Complexes," Allyn and Bacon, Boston, Mass., 1974.

(9) C. H. Langford and H. B. Gray, "Ligand Substitution Processes," W. A. Benjamin, New York, N.Y., 1965.

(10) J. F. Coetzee, in J. F. Coetzee and C. D. Ritchie, ed., "Solute-Solvent Interactions," Vol. 2, Marcel Dekker, New York (1976).

(11) K. Kustin, R. F. Pasternack, and E. M. Weinstock, J. Amer. Chem. Soc., 88, 4610 (1966).

(12) M. Eigen, Z. Phys. Chem. (Frankfurt am Main), 1, 176 (1954).

(13) R. M. Fuoss, J. Amer. Chem. Soc., 80, 5059 (1958).

(14) Chin-Tung Lin and D. B. Rorabacher, Inorg. Chem., 12, 2402 (1973).

(15) T. J. Swift and R. E. Connick, J. Chem. Phys., 37, 307 (1962).

(16)   R. E. Connick and D. Fiat, J. Chem. Phys., 44, 4103 (1966).

(17)   J. W. Neely and R. E. Connick, J. Amer. Chem. Soc., 94, 3419, 8646 (1972).

(18)   D. Rablen and G. Gordon, Inorg. Chem., 8, 395 (1969).

(19)   A. G. Desai, H. W. Dodgen and J. P. Hunt, J. Amer. Chem. Soc., 91, 5001 (1969).

(20)   Z. Luz and S. Meiboom, J. Chem. Phys., 40, 2686 (1964).

(21)   F. W. Breivogel, Jr., J. Phys. Chem., 73, 4203 (1969).

(22)   N. A. Matwiyoff and S. V. Hooker, Inorg. Chem., 6, 1127 (1967).

(23)   D. K. Ravage, T. R. Stengle and C. H. Langford, Inorg. Chem., 6, 1952 (1967).

(24)   J. F. O'Brien and W. L. Reynolds, Inorg. Chem., 6, 2110 (1967).

(25)   I. D. Campbell, J. P. Carber, R. A. Dwek, A. J. Nummelin, and R. E. Richards, Mol. Phys., 20, 913, 933 (1971).

(26)   R. J. West and S. F. Lincoln, Inorg. Chem., 11, 1688 (1972).

(27)   S. Thomas and W. L. Reynolds, J. Chem. Phys., 46, 4164 (1967).

(28)   S. Blackstaffe and R. D. Dweck, Mol. Phys., 15, 279 (1968).

(29)   N. S. Angerman and R. B. Jordan, Inorg. Chem., 8, 2579 (1969).

(30)   L. S. Frankel, Chem. Comm., 1254 (1969).

(31)   P. A. Cock, C. E. Cottrell, and R. K. Boyd, Can. J. Chem., 50, 402 (1972).

(32)   J. C. Boubel and J. J. Delpuech, Mol. Phys., 27, 113 (1974).

(33)   N. A. Matwiyoff, Inorg. Chem., 5, 788 (1966).

(34)   J. S. Babiec, C. H. Langford, and T. R. Stengle, Inorg. Chem., 5, 1362 (1966).

(35)   D. B. Rorabacher, Inorg. Chem., 5, 1891 (1966).

(36)   G. A. Melson and R. G. Wilkins, J. Chem. Soc., 4209 (1962).

(37)   R. H. Holyer, C. D. Hubbard, S.F.A. Kettle, and R. G. Wilkins, Inorg. Chem., 4, 929 (1965); 5, 622 (1966).

(38) P. K. Chattopadhyay and B. Kratochvil, Canad. J. Chem., 54, 2540 (1976).

(39) P. K. Chattopadhyay and B. Kratochvil, Inorg. Chem., 15, 3104 (1976).

(40) P. K. Chattopadhyay and J. F. Coetzee, Inorg. Chem., 12, 113 (1973).

(41) P. Ellis, R. Hogg, and R. G. Wilkins, J. Chem. Soc., 3308 (1959).

(42) F. Dickert, H. Hoffmann, and T. Janjic, Ber. Bunsenges. Phys. Chem., 78, 712 (1974).

(43) G. H. Nancollas and N. Sutin, Inorg. Chem., 3, 360 (1964).

(44) D. W. Margerum, D. B. Rorabacher, and J.F.G. Clarke, Jr., Inorg. Chem., 2, 667 (1963).

(45) R. G. Pearson and J. W. Moore, Inorg. Chem., 5, 1523 (1966).

(46) R. G. Pearson and O. P. Anderson, Inorg. Chem., 9, 39 (1970).

(47) R. B. Jordan, Inorg. Chem., 15, 748 (1976).

(48) W. J. MacKellar and D. B. Rorabacher, J. Amer. Chem. Soc., 93, 4379 (1971).

(49) J. F. Coetzee and G. R. Padmanabhan, J. Amer. Chem. Soc., 87, 5005 (1965).

(50) J. F. Coetzee and C. G. Karakatsanis, unpublished, preliminary results.

(51) P. Moore and D.M.W. Buck, J. Chem. Soc., Dalton, 1602 (1973).

(52) F. Dickert and H. Hoffmann, Ber. Bunsenges. Phys. Chem., 75, 1320 (1971).

(53) F. Dickert, H. Hoffmann, and W. Jaenicke, Ber. Bunsenges. Phys. Chem., 74, 500 (1970).

(54) H. Hoffmann, T. Janjic, and R. Sperati, Ber. Bunsenges, Phys. Chem., 78, 223 (1974).

(55) A. Diamond, A. Fanelli, and S. Petrucci, Inorg. Chem., 12, 611 (1973).

(56) H. P. Bennetto and E. F. Caldin, J. Chem. Soc. A, 2191, 2198,
      2207 (1971); H. P. Bennetto, ibid., 2211 (1971).

(57) J. A. Riddick and W. B. Bunger, "Organic Solvents," Wiley-
      Interscience, New York (1970).

(58) T. B. Reddy, Pure Appl. Chem., 25, 459 (1971).

(59) J. Juillard, Pure Appl. Chem., in press.

(60) J. F. Coetzee, Pure Appl. Chem., 13, 427 (1966).

(61) J. F. Coetzee, Pure Appl. Chem., in press.

(62) J. F. Coetzee and K. Umemoto, Inorg. Chem., 15, 3109 (1976).

(63) T. Fujinaga and K. Izutsu, Pure Appl. Chem., 27, 275 (1971).

(64) J. F. Coetzee and J. L. Hedrick, J. Phys. Chem., 67, 221
      (1963).

(65) P. K. Chattopadhyay and J. F. Coetzee, Inorg. Chem., 15, 400
      (1976).

(66) R. G. Pearson and P. Ellgen, Inorg. Chem., 6, 1379 (1967).

(67) E. F. Caldin and H. P. Bennetto, J. Solution Chem., 2, 217
      (1973).

(68) D.M.W. Buck and P. Moore, J. Chem. Soc. Dalton, 2082 (1974).

(69) H. P. Bennetto and Z. S. Imani, J. Chem. Soc. Faraday I, 71,
      1143 (1973).

(70) P. K. Chattopadhyay and J. F. Coetzee, Anal. Chem., 46, 2014
      (1974).

(71) J. F. Coetzee and E. J. Subak, Jr., to be published.

(72) J. F. Coetzee and D. M. Gilles, Inorg. Chem., 15, 405 (1976).

(73) M. L. Sanduja and W. M. Smith, Canad. J. Chem., 51, 3975
      (1973).

(74) W. M. Smith, private communication, 1975; preliminary value
      for pyridine.

(75) J. F. Coetzee and C. G. Karakatsanis, Inorg. Chem., 15, 3112
      (1976).

(76) J. F. Coetzee and D. Frollini, Jr., to be published.

(77) J. F. Coetzee and K. Umemoto, Inorg. Chem., 15, 3109 (1976).

(78) J. F. Coetzee, Pure Appl. Chem., in press.

(79) W. W. Brandt, F. P. Dwyer, and E. C. Gyarfas, Chem. Rev., 54, 959 (1954).

(80) W. R. McWhinnie and J. D. Miller, Adv. Inorg. Chem. Radiochem., 12, 135 (1969).

(81) K. Schofield, "Hetero-Aromatic Compounds, Pyrroles and Pyridines," Plenum Press, New York (1967).

(82) R. D. Nelson, Jr., D. R. Lide, Jr., and A. A. Maryoff, "Selected Values of Electric Dipole Moments for Molecules in the Gas Phase," NSRDS-NBS 10.

(83) P. E. Fielding and R.J.W. LeFevre, J. Chem. Soc., 1811 (1951).

(84) J. F. Coetzee and E. Hsu, J. Solution Chem., 4, 45 (1975).

(85) L. S. Frankel, Chem. Comm., 1254 (1969); Inorg. Chem., 10, 814 (1971).

(86) B. G. Cox, G. R. Hedwig, A. J. Parker, and D. W. Watts, Aust. J. Chem., 27, 477 (1974).

(87) C. H. Langford and J.P.K. Tong, Canad. J. Chem., 53, 702 (1975).

(88) J. S. Fritz, F. W. Cagle, Jr., and G. F. Smith, J. Amer. Chem. Soc., 71, 2480 (1949).

INFLUENCE OF THE SOLVENT ON THE ORIENTATION OF AMBIDENTATE SOLVENT

AND LIGAND MOLECULES IN THE COORDINATION SPHERE OF METAL IONS

Floyd Farha, Jr., Wayne C. Boring, Merton R. Olds, and
Reynold T. Iwamoto

Department of Chemistry, University of Kansas
Lawrence, Kansas 66045

Studies on the nature and behavior of inorganic species in
nonaqueous media, over the years, have dealt primarily with mono-
functional solvents such as acetonitrile, acetone, dimethyl
sulfoxide, dimethylformamide, and ethanol. In this paper, we
describe some of our work on the nature and behavior of inorganic
species in difunctional solvents where there is the possibility of
monocoordination by either or both functional groups. More specifically, our studies focus on the influence of the solvent on the
orientation of difunctional or ambidentate solvent molecules in
the coordination sphere of metal ions.

We shall examine first the nature of the copper(II) and
copper(I) species in the difunctional or ambidentate solvent hydracrylonitrile. This solvent molecule has a nitrile group and a
hydroxyl group, both capable of coordinating to copper ions. Our
interest in this study is then the mode of coordination of copper
(II) and of copper(I) ions by hydracrylonitrile. In beginning
the examination, we make the assumption that the properties of
hydracrylonitrile with regard to solvation of copper(II) and
copper(I) ions are a composite of those of ethanol and acetonitrile.

The potentials of copper couples in ethanol and in acetonitrile are: (1)

	Ethanol	Acetonitrile	Hypothetical system: Cu(I) solvated by acetonitrile; Cu(II) by ethanol
$E^{\circ\prime}_{Cu(II),Cu(I)}$ vs SCE	+0.12 V	+0.78 V	+0.70 V
$E^{\circ\prime}_{Cu(I),Cu}$ vs. SCE	+0.06 V	-0.52 V	-0.52 V

These potentials indicate that copper(I) ion is more strongly solvated in acetonitrile than in ethanol and copper(II) ion is slightly more strongly solvated in ethanol than in acetonitrile. The potentials for the copper couples in 1:1 ethanol-acetonitrile mixture support this information with regard to the solvation of copper(II) and copper(I) ions in ethanol and in acetonitrile.

	Acetonitrile	1:1 Ethanol-Acetonitrile Mixture
$E^{\circ\prime}_{Cu(II),Cu(I)}$ vs SCE	+0.78 V	+0.48 V
$E^{\circ\prime}_{Cu(I),Cu}$ vs SCE	-0.52 V	-0.46 V

The nearly identical negative potential of the copper(I), copper couple in 1:1 ethanol-acetonitrile compared with that in acetonitrile points to acetonitrile coordination of copper(I) ion in the mixture, and the more negative potential of the copper(II), copper(I) couple in the mixture compared to that in acetonitrile is consistent with ethanol coordination of copper(II) ion.

In hydracrylonitrile, copper(II) ion, therefore, would be expected to be coordinated by the hydroxyl group and copper(I) ion by the nitrile group. The potentials of the copper couples in hydracrylonitrile,

$$E^{\circ\prime}_{Cu(II),Cu(I)} \text{ vs SCE } = +0.78 \text{ V}$$

$$E^{\circ\prime}_{Cu(I),Cu} \text{ vs SCE } = -0.46 \text{ V}$$

however, indicate that copper(II) ion is not coordinated by the hydroxyl group, as expected, but like copper(I) ion, by the nitrile group.

Importantly, this unexpected electrochemical picture of the solvation of copper(II) ion in hydracrylonitrile is supported by infrared and visible spectral data (Table I). When copper(II)

Table I. Visible and Infrared Data of Acetonitrile, Ethanol, 1:1 Ethanol-Acetonitrile, and Hydracrylonitrile Solutions Containing Copper(II) Perchlorate.

	Visible Data λ_{max} (nm)	Infrared Data			
		OH (cm^{-1})		CN (cm^{-1})	
		solv.	new band	solv.	new band
Acetonitrile	760-770			2254	
Acetonitrile + Cu^{++}					2299
EtOH		3420 (br.)			
EtOH + Cu^{++}	815		br. sh. on lower freq. side		
1:1 EtOH-Acetonitrile		3420 (br.)		2257	
1:1 EtOH-Acetonitrile + Cu^{++}	815		br. sh. on lower freq. side		no change
Hydracrylonitrile		3420 (br.)		2250	
Hydracrylonitrile + Cu^{++}	760-770		no change		2300

perchlorate is added to acetonitrile a new band for coordinated
nitrile appears at 2299 cm^{-1} in the infrared spectrum, and a band
appears in the visible region at 760-770 nm. Addition of copper
(II) perchlorate to ethanol results in the appearance of a broad
shoulder on the lower frequency side of the 3420 cm^{-1} OH band and
a band in the visible region at 815 nm. When copper(II) perchlorate is introduced into 1:1 ethanol-acetonitrile mixture, no new
band appears in the 2300 cm^{-1} region for coordinated nitrile.
There is a broad shoulder on the lower frequency side of the 3420
cm^{-1} OH band, and the visible spectrum shows a band at 815 nm.
This infrared and visible spectral information supports the electrochemical data that in 1:1 ethanol-acetonitrile mixture cop-
per(II) ion is solvated by ethanol molecules. In the case of
hydracrylonitrile, the addition of copper(II) perchlorate results
in the appearance in the infrared spectrum of a new band for
coordinated nitrile at 2300 cm^{-1} and in the visible spectrum a
band at 760-770 nm. No change is observed in the infrared spectrum
in the 3420 cm^{-1} region. These data clearly indicate solvation of
copper(II) ion by hydracrylonitrile through the nitrile group.

The unexpected mode of solvation of copper(II) ion in hydracrylonitrile appears to be due to the fact that of the two
possible solvated species, one with nitrile groups coordinated to
copper(II) ion and the other with hydroxyl groups attached to
copper(II) ion, the one that is more polar and therefore more
compatible with the polar hydracrylonitrile medium is the former
where there are hydroxyl groups in the outer sheath of the solvated species. In 1:1 ethanol-acetonitrile mixture, also a polar
medium, the favored solvated species would be the more polar
ethanol solvate. The acetonitrile solvate with methyl groups in
the outer sheath would be distinctly nonpolar.

Consistent with this explanation, visible spectral studies of
nitromethane solutions containing hydracrylonitrile (Table II) show
that in this polar medium nitrile-group coordination of copper(II)
ion by hydracrylonitrile molecules is favored, yielding the more
polar of the two possible coordinated species, and in solutions
containing equimolar amounts of ethanol and acetonitrile, coordination by ethanol molecules is favored. Similar studies of solutions of 1,2-dichloroethane containing hydracrylonitrile indicate
that in these solutions of low polar character copper(II) ion is
coordinated by hydracrylonitrile molecules through the hydroxyl
group, yielding the less polar solvated species, while in solutions
containing equimolar amounts of ethanol and acetonitrile, partial
coordination by acetonitrile is suggested. It is clear that in
hydracrylonitrile the compatibility of the solvated species with
the solvent plays an important role in determining which functional group, the hydroxyl or the nitrile, coordinates to a metal
ion such as copper(II) ion and which group forms the outer sheath
of the solvated species.

Table II. Visible Spectral Data for Solutions of Copper(II)

Perchlorate in Nitromethane and 1,2-Dichloroethane

Containing Ethanol, Acetonitrile, Hydracrylonitrile,

and Equimolar Amounts of Ethanol and Acetonitrile

Medium	λ_{max} (nm)
Nitromethane, 1$\underline{M}$ ethanol	805
Nitromethane, 1$\underline{M}$ acetonitrile	770
Nitromethane, 0.5$\underline{M}$ hydracrylonitrile	780
Nitromethane, 0.25$\underline{M}$ ethanol and 0.25$\underline{M}$ acetonitrile	800
1,2-Dichloroethane, 1$\underline{M}$ ethanol	805
1,2-Dichloroethane, 1$\underline{M}$ acetonitrile	780
1,2-Dichloroethane, 0.5$\underline{M}$ hydracrylonitrile	805
1,2-Dichloroethane, 1$\underline{M}$ ethanol and 1$\underline{M}$ acetonitrile	795

Table III. Summary of Electrochemical, Infrared, and Proton Magnetic Resonance Data on 3-Butenenitrile, 30% 3-Butenenitrile--Propylene Carbonate, and 30% 3-Dimethylaminopropionitrile--1,2-Dichloroethane Solutions Containing Copper(I) and Silver Ions.

Electrochemical

	$E^{\circ\prime}_{Cu(I),Cu(Hg)}$ [a]	$E^{\circ}_{Ag(I),Ag(Hg)}$ [a]
3-Butenenitrile	-0.26 V	+0.47 V
Propionitrile	-0.27 V	+0.47 V
30% 3-Butenenitrile--Propylene Carbonate	+0.11 V	>+0.66 V
30% Propionitrile--Propylene Carbonate	-0.13 V	+0.62 V
30% 3-Dimethylaminopropionitrile--1,2-Dichloroethane	-0.24 V[b]	+0.32 V[c]
30% 3-Dimethylaminopropionitrile--Propylene Carbonate	-0.27 V[b]	+0.28 V[c]
30% Propionitrile--1,2-Dichloroethane	-0.20 V[b]	+0.51 V[c]

Infrared

	$\nu_{C=C}$	$\nu_{C\equiv N}$
3-Butenenitrile, 0.5 F MClO$_4$	1625 and 1635 cm^{-1} no change	2220 and 2250 cm^{-1} new bands 2238 and 2268 cm^{-1}
30% 3-Butenenitrile--Propylene Carbonate, 0.5 F AgClO$_4$	new band 1565 cm^{-1}	new bands 2238 and 2268 cm^{-1}

Table III (Continued)

	ν_{C-N} (asymmetric) 1010 cm^{-1}	$\nu_{C\equiv N}$ 2245 cm^{-1}
30% 3-Dimethylaminopropionitrile-- 1,2-Dichloroethane, 0.5 F AgClO$_4$	new band 980 cm^{-1}	new band 2255 cm^{-1}

Proton magnetic resonance

	H$_2$C=CH (5.5 ppm)^d	CH$_2$ (3.2 ppm)^d
3-Butenenitrile, 0.5 F AgClO$_4$	no change	downfield shift 0.05 ppm
30% 3-Butenenitrile-- Propylene Carbonate, 0.5 F AgClO$_4$	downfield shift 0.23 ppm	downfield shift 0.08 ppm
30% 3-Butenenitrile-- 1,2-Dichloroethane, 0.5 F AgClO$_4$	no change	downfield shift 0.13 ppm

	CH$_3$ (2.3 ppm)	CH$_2$ (2.5 ppm)
30% 3-Dimethylaminopropionitrile-- 1,2-Dichloroethane, 0.5 F AgClO$_4$	downfield shift 0.01 ppm	downfield shift 0.05 ppm

[a] E values are vs. SCE
[b] E°Cu(I),Cu
[c] E°Ag(I),Ag
[d] complicated spectrum - only major peak reported

To reinforce this point, the influence of the solvent on the
mode of coordination of 3-dimethylaminopropionitrile and that of
3-butenenitrile to silver and copper(I) ions has been examined.(2)
3-Butenenitrile has an olefinic group and a nitrile group both
capable of coordinating to silver and copper(I) ions. Because the
two groups do not differ markedly in their abilities to coordinate
to the two metal ions and because the nitrile group is polar and
the olefinic group nonpolar, in a polar medium we would expect
olefinic coordination of the metal ions with nitrile groups in the
outer sheath of the solvated species, and in a nonpolar medium,
nitrile coordination with olefinic groups in the outer sheath of
the solvated species.

Electrochemical, infrared, and proton nuclear magnetic resonance studies show that in 3-butenenitrile and in the less polar
medium 30% 3-butenenitrile--1,2-dichloroethane there is complete
nitrile coordination of silver and copper(I) ions. Data for the
case of 3-butenenitrile solutions are presented in Table III. The
identical potential of each of the metal ion (I), metal couples
with that for the corresponding couple in propionitrile, in which
there can only be nitrile coordination, clearly establishes
coordination of copper(I) and silver ions by nitrile groups in
3-butenenitrile. In addition, the infrared and proton magnetic
resonance data shown indicate no olefinic coordination of silver
ion. In the polar mixture 30% 3-butenenitrile--propylene carbonate,
similar experimental measurements (Table III) suggest that there
is a good deal of olefinic coordination of the two metal ions, but
complete olefinic coordination with the polar nitrile groups in the
outer sheath of the solvated species is not possible because of the
somewhat greater coordinating ability of the nitrile group than of
the olefinic group for the metal ions.

In 3-dimethylaminopropionitrile and in the polar 30% 3-dimethyl-
aminopropionitrile-propylene carbonate mixture, electrochemical
infrared, and proton magnetic resonance studies indicate extensive
tertiary amine coordination of the two metal ions in the former
solution and complete tertiary amine coordination in the latter.
In the mixture 30% 3-dimethylaminopropionitrile--1,2-dichloroe-
thane, both nitrile and tertiary amine coordination of copper(I)
and silver ions are evident. Data for this solution of low polar
character are shown in Table III. The potentials of the copper(I),
copper and silver(I), silver couples in this solution are between
those in the propylene carbonate solution, in which there is
complete tertiary amine group coordination of the metal ions, and
those in propionitrile--1,2-dichloroethane solutions, in which
there is only nitrile coordination. The infrared data for solutions of silver ion indicate both tertiary amine and nitrile
coordination of the metal ion, but, importantly, the intensity of
the band resulting from coordinated tertiary amine group is quite

weak.   The proton magnetic resonance data also reflect only limited extent of tertiary amine group coordination of silver ion in this medium.

To determine whether a difunctional molecule such as 4-cyano-phenylacetonitrile with the same coordinating group (nitrile) in two different environments would exhibit the same solution phenomenon shown by hydracrylonitrile, 3-butenenitrile and 3-dimethylaminopropionitrile, (3) the coordination of this dinitrile to silver ion in propylene carbonate and in nitromethane was examined.   One of the nitrile groups of 4-cyanophenylacetonitrile is attached to an alkyl group (aliphatic nitrile) and the other to a benzene ring (aromatic nitrile).   The properties of 4-cyano-phenylacetonitrile, it would appear, should be a composite of those of benzonitrile and phenylacetonitrile.   The dipole moments 4.1 and 3.5 debye units for benzonitrile and phenylacetonitrile, respectively, suggest that the aromatic nitrile end of 4-cyano-phenylacetonitrile is more polar than the aliphatic nitrile end. The basicity and the donor number (coordinating ability) of phenylacetonitrile have been reported to be greater than those of benzonitrile. (4, 5)   All three of these features favor in a polar medium aliphatic nitrile coordination of silver ion, the more basic and better donor coordinated to silver ion and the more polar nitrile in the outer sheath of the coordinated species. Quite surprisingly, the nitrile stretching bands in the infrared for propylene carbonate solutions containing silver ion and 4-cyanophenylacetonitrile revealed that in this polar medium there was essentially complete aromatic nitrile coordination of silver ion.   No evidence for coordinated aliphatic nitrile group was found.   This mode of coordination of silver ion in a polar medium is completely contrary to that expected on the basis of the charac-teristics of benzonitrile and phenylacetonitrile just described. In the case of the less polar solvent nitromethane, there appears to be, as expected, about 50% aromatic nitrile coordination of silver ion by 4-cyanophenylacetonitrile.

To explain the unexpected mode of coordination of silver ion by the dinitrile in propylene carbonate, the formation constants of the benzonitrile and phenylacetonitrile complexes of silver ion in propylene carbonate and in nitromethane were examined. These constants are:

Ligand	Nitromethane Solutions		Propylene Carbonate Solutions	
	K_{f1}	K_{f2}	K_{f1}	K_{f2}
Phenylace-tonitrile	88 ± 10	13 ± 2	7.2 ± 2.5	12.6 ± 3.2
Benzonitrile	66 ± 10	9 ± 3	32 ± 3	5 ± 1

The order of stability of the benzonitrile and phenylacetonitrile complexes of silver ion in nitromethane solutions is consistent with the properties of benzonitrile and phenylacetonitrile. For propylene carbonate solutions, however, the formation constants for the benzonitrile and phenylacetonitrile complexes of silver ion are in the reverse order from that expected. In addition, the small formation constant for the phenylacetonitrile complex of silver ion in propylene carbonate suggests that this nitrile, the aliphatic nitrile group, is unexpectedly strongly solvated in propylene carbonate.

From the formation constants, the following estimates of the extent of aromatic nitrile vs aliphatic nitrile coordination of silver ion in solutions containing equimolar amounts of benzonitrile and phenylacetonitrile were obtained: For propylene carbonate solutions, ca 70% aromatic nitrile coordination, and for nitromethane solutions, ca 40%. There is then a difference of 30% in the extent of aromatic nitrile coordination observed with 4-cyanophenylacetonitrile and that estimated for an equimolar mixture of benzonitrile and phenylacetonitrile in propylene carbonate and 10% in the case of nitromethane solutions. These differences arise undoubtedly from the stronger solvation of the aliphatic nitrile group than of the aromatic nitrile group by these solvents. Such solute-solvent interaction would promote aromatic nitrile coordination of silver ion by 4-cyanophenylacetonitrile.

Finally, a brief account of a related study on the influence of the solvent on the mode of protonation of 2,2'-dipicolylamine is presented. (6) Of particular interest is the effect of solvent on the nature of the diprotonated 2,2'-dipicolylamine species. Proton magnetic resonance study of the protonation of 2,2'-dipicolylamine in water (D_2O), acetonitrile, and benzonitrile yielded the chemical shift data in Figures 1 and 2. There is the expected downfield shift of the signal of the methylene hydrogens on the addition of the first proton to the tribasic amine, consistent with protonation of the aliphatic amine, the strongest base. On addition of the second proton on the tribasic amine, an upfield shift of the signal of the methylene hydrogens occurs, suggesting deprotonation of the aliphatic amine. With the addition of the third proton on the amine, once again there is a downfield shift of the signal of the methylene hydrogens. Figure 2 provides information on the protonation of the pyridine ring. As expected from the chemical shift data of the methylene hydrogens, there is no change in the position of the signal of the ring hydrogen in the 5 position during the addition of the first proton on the amine. On addition of the second proton, there is a large downfield shift which has been established as reflecting full protonation of both pyridine rings. The addition of the third proton on

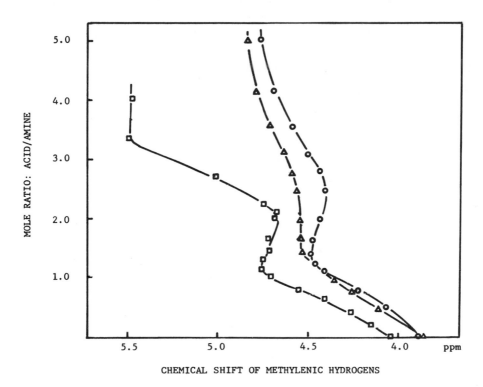

CHEMICAL SHIFT OF METHYLENIC HYDROGENS

Figure 1. Shift in the proton magnetic resonance signal of the methylenic hydrogens of 2,2'-dipicolylamine on protonation of the amine in:

(a) - △ - water (D_2O) with 70% perchloric acid. Chemical shift vs. 3-(trimethylsilyl)propane sulfonate.

(b) - ○ - acetonitrile with 70% perchloric acid. Chemical shift vs. tetramethylsilane.

(c) - □ - benzonitrile with trifluoromethane sulfonic acid. Chemical shift vs. tetramethylsilane.

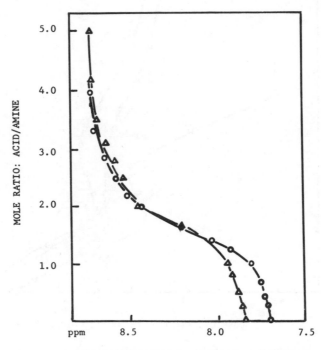

CHEMICAL SHIFT OF RING HYDROGEN - 5 POSITION

Figure 2.   Shift in the proton magnetic resonance signal of
the ring hydrogen in the 5-position of 2,2'-dipicolylamine on
protonation of the amine in:

(a)   - △ -   water(D_2O) with 70% perchloric acid.   Chemical
              shift vs. 3-(trimethylsilyl)propane sulfonate.

(b)   - ◯ -   acetonitrile with trifluoromethane sulfonic
              acid.   Chemical shift vs. tetramethylsilane.

the amine results in only a small change in the position of the signal of the ring hydrogen, a behavior consistent with protonation of the aliphatic amine. The chemical shift patterns for the methylene hydrogens and the ring hydrogen in the 5 positions in 2,2'-dipicolylamine on protonation of the amine in water (D$_2$O), acetonitrile, and benzonitrile can be accounted for as follows:

(a)

$$+HPyCH_2NCH_2PyH^+$$

$$PyCH_2NCH_2Py \xrightarrow{H^+} PyCH_2NCH_2Py \longrightarrow \quad (b) \qquad \xrightarrow{H+} \quad +HPyCH_2NCH_2PyH^+$$

$$PyCH_2NCH_2PyH^+$$

A quantitative study of the influence of the solvent on the distribution of the two diprotonated species using substituent shielding constants for the methylenic group indicates that in water (D$_2$O) 56% of the diprotonated species is the electrostatic- ally less favorable (higher energy) species (b). The percentage of this species drops to 32% in less polar acetonitrile, and still lower to 21% in much less polar benzonitrile. The more polar the medium the greater is the ability of the medium to accommodate the electrostatically less favored diprotonated form with protons on adjacent basic sites.

SUMMARY

The solvation of copper(II) ion in hydracrylonitrile and in 1:1 ethanol-acetonitrile mixture was examined by electrochemical and infrared and visible spectral techniques. In hydracryloni- trile, nitrile coordination of copper (II) ion was found to be favored, while in 1:1 ethanol-acetonitrile mixture, ethanol coordination of copper(II) ion was indicated. In solvents which contain a polar and a nonpolar coordinating group with nearly equal abilities for coordinating to a metal ion, the compatibility of the solvated species with the solvent appears to be an impor- tant factor in determining which group coordinates to the metal ion and which group makes up the outer sheath of the solvated species. Studies on the influence of the solvent on the orienta- tion of 3-butenenitrile and 3-dimethylaminopropionitrile in the coordination sphere of copper(I) and silver ion reinforce this solvation phenomenon. In another study, the mode of coordination of the dinitrile 4-cyanophenylacetonitrile to silver ion in propy- lene carbonate and in nitromethane was examined. In propylene carbonate, aromatic nitrile coordination was found to be the only mode of nitrile coordination of silver ion, contrary to that expected on the basis of the properties of benzonitrile and phenylacetonitrile. In nitromethane, aromatic nitrile coordination,

as expected, accounts for about one-half of the nitrile coordination of silver ion. These results have been explained through the
stability constants of the benzonitrile and phenylacetonitrile
complexes of silver ion in the two solvents. Finally, a proton
magnetic resonance study of the influence of the solvents water
(D_2O), acetonitrile, and benzonitrile on the protonation scheme of
2,2'-dipicolylamine is described.

ACKNOWLEDGEMENT

     Support of these studies by the National Science Foundation;
the Directorate of Chemical Sciences, Air Force Office of Scientific Research; The Petroleum Research Fund, administered by the
American Chemical Society; DuPont Chemical Company; and the General
Research Fund, University of Kansas, is gratefully acknowledged.

REFERENCES

(1) F. Farha, Jr. and R. T. Iwamoto, J. Electroanal. Chem., 8, 55
     (1964).

(2) F. Farha, Jr. and R. T. Iwamoto, ibid., 13, 390 (1967).

(3) W. C. Boring and R. T. Iwamoto, Inorg. Chim. Acta, 7, 264
     (1973).

(4) J. F. Coetzee and D. K. McGuire, J. Phys. Chem., 67, 1810
     (1963).

(5) V. Gutman, "Coordination Chemistry in Nonaqueous Solutions,"
     Springer-Verlag, New York, New York, 1968, p. 19.

(6) M. R. Olds and R. T. Iwamoto, Anal. Chem., 47, 2394 (1975).

ALKALI METAL, MAGNESIUM-25, AND SILVER-109 NMR STUDIES OF COMPLEX COMPOUNDS IN NONAQUEOUS SOLVENTS

Alexander I. Popov

Department of Chemistry, Michigan State University

East Lansing, Michigan   48824

INTRODUCTION

At the present time we have at our disposal a large variety of experimental techniques for the characterization of solute species in solutions and the studies of their properties. Among these nuclear magnetic resonance of metallic nuclei is becoming an increasingly important probe of their behavior in solutions and, in particular, of very weak interactions of the respective ions with solvents, counterions, or complexing agents. This presentation will attempt to describe briefly some of the recent applications of alkali metal, magnesium-25, and silver-109 nmr measurements to the study of ions in solutions.

Table I illustrates the important properties of the above nuclei. Except for silver, all nuclei have spins of greater than 1/2 and, therefore, a quadrupole moment. Nevertheless, except for rubidium-85 and rubidium-87, the resonance linewidths are reasonably narrow and in the case of lithium-7 and cesium-133 resonances, the natural linewidth is less than 1 Hz.

Both of the naturally occuring silver nuclei have spin of 1/2. The relaxation time is extremely long; T_1 for silver-109 is approximately 50 sec.(1) It is a common practice to shorten the relaxation time by doping the solutions with small amounts of paramagnetic ions.

As seen from Table I, the sensitivities of the nuclei vary within large limits and are much below that of the proton. In particular the sensitivities of potassium and silver nuclei are quite low. However, the recent advent of Fourier transform

TABLE I

Nuclear Properties of Alkali, Magnesium and Silver Nuclei

Isotope	NMR Frequency at 14.1 kgauss (MHz)	% Natural Abundance	Sensitivity at Constant Field vs ^{1}H	Spin
^{6}Li	8.834	7.43	0.00851	1
^{7}Li	23.331	92.57	0.294	3/2
^{22}Na	6.252	0.0	0.0181	3
^{23}Na	15.879	100	0.0927	3/2
^{39}K	2.802	93.09	0.00051	3/2
^{40}K	3.483	0.012	0.00521	4
^{41}K	1.540	6.41	0.000084	3/2
^{85}Rb	5.797	72.8	0.0105	5/2
^{87}Rb	19.644	27.8	0.177	3/2
^{133}Cs	7.875	100	0.0474	7/2
^{25}Mg	3.751	10.05	0.0268	5/2
^{107}Ag	2.428	51.35	0.000067	1/2
^{109}Ag	2.793	48.65	0.00010	1/2

spectrometers and of high field superconductive solenoids has increased ·the sensitivity of nmr measurements by several orders of magnitude. At this time it is not uncommon to measure resonances of 10^{-2} - 10^{-4} M salt solutions.

One of the first studies of sodium-23 nmr in aqueous solutions was done by Wertz and Jardetsky in 1956.[2] The authors measured the sodium-23 line width and line amplitude of numerous inorganic salts. The work was later extended to a qualitative study of sodium complexes with organic and inorganic anions.[3]

Chemical shifts of sodium-23, potassium-39, rubidium-87, and cesium-133 in aqueous solutions were measured by Richards and coworkers.[4,5] Salt concentration and the nature of the counter-ion strongly influence the magnitudes of the chemical shifts. In general, the cationic chemical shifts are downfield from the standard (respective alkali ion in aqueous solution at infinite dilution) for the halides, while polyatomic oxyanions shift the alkali ion resonance upfield.

Bloor and Kidd studied sodium-23 magnetic resonance of sodium iodide solutions in fourteen nonaqueous solvents.[6] The sodium-23 chemical shifts in the nonaqueous solutions were referred to an aqueous sodium iodide solution at infinite dilution.

The authors postulate that the sodium-23 chemical shifts are strongly influenced by the effect of the overlapping of the p-orbitals of the sodium ion with the s and p-orbitals of the neighboring species (solvent molecules or other ions) on the para-magnetic screening constant σp. Thus the sodium-23 resonance would be at a lower field in a more basic solvent. A comparison of the relative chemical shifts with the acidity constants of these solvents in aqueous solutions does indeed, indicate some correlation between the two parameters. It appears doubtful, however, that the authors' assumption of a linear correlation is valid.

Our sodium-23 nmr study of several sodium salts in a number of nonaqueous solvents showed that the nature of the anion has a strong influence on the resulting chemical shift.[7,8] The measurements were made vs aqueous saturated sodium chloride solution and corrected for the bulk susceptibility differences between the samples and the reference solution.

The sodium-23 resonance in perchlorate and tetraphenylborate salts is essentially concentration independent, but there is a large concentration dependence for the halides and the thio-cyanates. It seems reasonable to assume that the change in the chemical shift is indicative of contact ion pair formation. In

each solvent, the chemical shifts converge at infinite dilution
to a limiting value which is characteristic of the free solvated
sodium ion in the particular solvent.

The distribution of sodium-23 chemical shifts in various
solvents indicated some correlation with the donor or the solvating
ability of the solvent.   The donor ability of a solvent, however,
is a somewhat nebulous factor which, at this time, cannot be
defined on the basis of theoretical considerations.   Several
empirical donor scales have been proposed by various authors.   The
choice of the scale is somewhat a matter of personal preference.
To this author it seems that the Gutmann "donicity" scale (9) is
by far the most useful.

Gutmann defines solvent donicity as the enthalpy of formation
of an antimony(V) chloride complex with the solvent molecule S,
the reaction taking place in an inert medium such as 1,2-
dichloroethane,

$$S + SbCl_5 \underset{\longleftarrow}{\overset{1,2\text{-DCE}}{\longrightarrow}} S \cdot SbCl_5$$

Donicity, therefore, is expressed as the negative value of enthalpy
(in kcal mole^{-1}) for the preceding reaction.

A plot of the limiting chemical shifts vs Gutmann's donicity
values resulted in a rather respectable straight line.(8)   The
only discrepancy seems to arise in aqueous solutions where the
plot predicts a value of 33 while the donicity value is 18.

The reasons for the linear relationship between the two
parameters are not immediately obvious.   In the case of the sodium
ion solvation, we have predominantly electrostatic ion-dipole
interactions while the donicity value reflects a donor-acceptor
interaction.

The fact that alkali metal nmr measurements are very sensitive probes for contact ion pair formation, helps to understand
the role of the donor properties of solvents in this process.
A comparison of ionic association in dimethyl sulfoxide and
sulfolane solutions illustrates this point.   While dimethyl
sulfoxide and sulfolane have nearly identical dielectric constants
of 46.7 and 44, their donor numbers of 29.8 and 14.9 respectively
indicate that dimethyl sulfoxide is a better solvating solvent.
Indeed, the sodium-23 chemical shift of sodium iodide solutions
in the former solvent shows no concentration dependence, while
there is a strong concentration dependence in sulfolane solutions.
Strong concentration dependence of the sodium-23 chemical shift

is also observed in propylene carbonate solutions of sodium iodide (10) which has a high dielectric constant of 65 but a relatively low donor number of 15.1. Therefore, Na^+I^- ion pairs are readily formed in sulfolane and propylene carbonate, but not in dimethyl sulfoxide.

LITHIUM-7 NMR OF LITHIUM SALTS IN VARIOUS SOLVENTS

Lithium-7 chemical shifts were observed and studied in several solvents by Maciel, et al. (11) and by Akitt and Downs.(12) Their data clearly indicated that the nature of the solvent does influence the lithium-7 relative chemical shift and that, particularly in nonaqueous solvents, chemical shift also varies with the salt concentration and with the nature of the counterion.

It is evident from these data that there is a definite counterion dependence of the chemical shift in "nonpolar" solvents. For example, in acetonitrile, dimethoxyethane and tetrahydrofuran, the lithium bromide chemical shifts are downfield from those observed in lithium perchlorate solution. On the other hand, in "polar" solvents - water, methanol and dimethyl sulfoxide, lithium-7 chemical shifts are anion-independent. The data indicate that in the former case the chemical shift is influenced by cation-anion interaction, presumably by the formation of contact ion pairs.

The solvent dependence of the chemical shift could not be correlated with physicochemical properties of solvent molecules such as the dielectric constant, the dipole moment or the heat of vaporization. The chemical shifts, however, seem to be influenced by the chemical nature of the solvent. For example, chemical shifts in carbonyl-containing solvents are grouped together indicating a specific influence resulting from coordination of the Li^+ ion with the carbonyl oxygen. Analogous explanations can be offered for the opposite effects of the lithium-7 chemical shifts in acetonitrile and in pyridine.(11)

We studied the chemical shift of lithium-7 resonance in eleven nonaqueous solvents.(13) The range of chemical shifts, extrapolated to infinite dilution, varied from +2.80 ppm in acetonitrile to -2.54 ppm in pyridine.

In dimethylformamide and, to a somewhat lesser extent, in propylene carbonate, methanol and dimethyl sulfoxide, the lithium-7 chemical shifts are essentially independent of the counterion and of the salt concentration. The results indicate that the salts are either largely dissociated or that they exist in the form of solvent-separated ion pairs. In tetrahydrofuran solutions the chemical shifts are essentially concentration-independent, but

they do depend on the nature of the counterion. This solvent has
a low dielectric constant of 7.58 but it is a reasonably good
solvating agent as shown by its Gutmann donor number of 20.0.(9)
Thus in tetrahydrofuran solutions the conditions would be favorable
for an equilibrium between solvent-separated and contact ion pairs.
It can be assumed with reasonable certainty, that in the concentration range studied (0.01 - 0.6 M) the salt exists predominantly
in the form of contact ion pairs and only at concentrations below
0.01 M the formation of solvent-separated ion pairs or free ions,
would be evident.

On the other hand, the data in acetic acid solutions look
very much like those in a strongly solvating polar solvent,
i.e. dimethylformamide or dimethyl sulfoxide. The chemical shifts
are essentially independent of the concentration and of the counterion. Yet acetic acid has a very low dielectric constant of 6.3
and it is obvious that in this medium, even in dilute solutions,
salts must exist as ion pairs. However, since there is no
concentration or counterion effect on the chemical shifts, it
seems reasonable to assume that the predominant solute species
are the solvent-separated ion pairs.

It is interesting to note that contrary to the sodium-23
chemical shifts in various solvents, the lithium-7 limiting
chemical shifts show no correlation with the Gutmann donor numbers.
In the case of the lithium-7 nucleus, however, the diamagnetic
and paramagnetic screening constants are of the same order of
magnitude and tend to cancel one another.(14) Therefore, effects
such as ring currents and neighbor-anisotropy effects become more
important for the lithium-7 chemical shifts.

CESIUM-133 CHEMICAL SHIFTS IN VARIOUS SOLVENTS

Although cesium-133 nucleus has a spin of 7/2, the electrical
quadrupole moment is small (-3 x 10^{-3} barns) and consequently the
resonance has a very narrow natural line width of < 1 Hz.

Concentration dependence of cesium-133 chemical shifts were
studied in water and in nine nonaqueous solvents.(15,16) Cesium
chloride, bromide, iodide, thiocyanate, perchlorate and tetraphenylborate were used. Solubility limitations were severe and
in some cases salts were essentially insoluble in the solvents of
interest.

As before, chemical shifts vary with concentration and with
the anion. Cesium-133 infinite dilution chemical shifts in various
solvents were obtained and are given in Table II. It is interesting
to note that there is a general tendency for the resonance to move

TABLE II

Cesium-133 Infinite Dilution Chemical Shifts
In Various Solvents (15,16)

Solvent	Donor Number	Dielectric Constant	Chemical Shift
Nitromethane	2.7	36.87	59.8 ± 0.2
Acetonitrile	14.1	37.5	-32.0 ± 0.4
Propylene Carbonate	15.1	69.0	35.2 ± 0.2
Acetone	17.0	20.7	26.8 ± 0.3
Formamide	24.7	11.0	27.9 ± 0.3
Methanol	25.7	32.7	45.2 ± 0.2
Dimethylformamide	26.6	36.7	0.5 ± 0.2
Dimethyl sulfoxide	29.8	46.7	-68.0 ± 0.2
Water	33.0	78.39	0.0
Pyridine	33.1	12.4	-29.4 ± 1

TABLE III

Ion Pair Formation Constants for $Cs^+Ph_4B^-$
In Various Solvents (16)

Solvent	Dielectric Constant	Donor Number	K_{ip}°
Pyridine	12.4	33.1	$(3.7 \pm 0.2) \times 10^2$
Acetonitrile	37.5	14.1	40 ± 10
Acetone	20.7	17.0	22 ± 3
Dimethylformamide	36.7	26.6	∿ 0
Dimethyl sulfoxide	46.7	29.8	∿ 0

downfield with increasing donicity of the solvent. In fact, only
the chemical shifts in three solvents, methanol, acetonitrile and
dimethyl sulfoxide, deviate significantly from the straight line.
Precisely the same behavior was observed in the case of potassium-39
nmr measurements (see below).

The variation of the cesium-133 chemical shift with the salt
concentration is related to the ion pair association constant.(16)
From the analysis of the data formation constants of cesium
tetraphenylborate ion pairs were calculated in several solvents.
The results are given in Table III.

It should be noted that in general, electrical conductance
measurements do not distinguish between different types of ion
pairs (tight or solvent-separated) while nmr measurements presumably give the formation constant for contact ion pairs. Thus
it seems feasible, in principle, to study free ions $\leftrightarrow$ contact
ion pairs $\leftrightarrow$ solvent-separated ion pairs equilibria by a combination of nmr and electrochemical techniques.

POTASSIUM-29, MAGNESIUM-25 AND SILVER-109 NMR STUDIES

The concentration, the anion and the solvent dependence of the
potassium-39 chemical shifts in general follow the same pattern as
for the other alkali cations. The infinite dilution chemical
shifts follow the same pattern as for cesium-133. The data are
given in Table IV.(17) A plot of δ_O vs Gutmann donor number gives
a reasonable straight line with the exception of methanol,
acetonitrile and dimethyl sulfoxide, that is, the same three
solvents which deviate from straight line in the cesium-133
donor number plots.

Previous studies of magnesium-25 and silver-109 resonances
are very sparse probably due to the low frequency of the two
resonances (magnesium-25 at 3.7 MHz and silver-109 at 2.8 MHz
both at 1.4 Tesla) and relatively low sensitivity. A thorough
investigation of magnesium-25 in aqueous solutions was recently
reported by Simeral and Maciel.(18) Measurements of the
magnesium-25 chemical shifts for magnesium sulfate, perchlorate,
nitrate, chloride and bromide solutions generally in the 0.05 -
4.1 $\underline{M}$ concentration range, showed that the shifts were only very
slightly concentration-dependent and that increased line width
with concentration could be ascribed to an increase in viscosity
in all cases except those of the chloride and the sulfate where
some of the line broadening could be due to cation-anion
interactions.

TABLE IV

Limiting Chemical Shift of the $Cs^+ \cdot (18C6)_2$ Complex
and the Thermodynamic Parameters for the Reaction
$Cs^+ \cdot 18C6 + 18C6 \longleftrightarrow Cs^+(18C6)_2$ in Pyridine

Temperature (°K)	$K_{1(min)}$ (M^{-1})	K_2 (M^{-1})	δ_2 (ppm)
297	10^5	79 ± 2	47.8
285	10^6	121 ± 5	49.4
272	10^6	218 ± 14	49.9
255	10^6	432 ± 58	51.4
244	10^6	623 ± 35	51.9
235	5×10^6	1173 ± 160	51.2

$$(\Delta G_2^\circ)_{298} = -2.58 \pm 0.02 \text{ kcal mole}^{-1}$$
$$\Delta H_2^\circ = -5.8 \pm 0.2 \text{ kcal mole}^{-1}$$
$$\Delta S_2^\circ = -10.7 \pm 0.6 \text{ cal mole}^{-1} \text{ deg}^{-1}$$

Our preliminary measurements of magnesium-25 resonances (at natural abundance) in aqueous solutions showed good agreement with those reported by Simeral and Maciel. The chemical shifts are essentially independent of salt concentration which indicates that the immediate environment of the Mg^{2+} ion in aqueous solutions remains essentially constant as the concentration of the salt is increased, i.e. the ion is strongly hydrated and even in concentrated solutions there are no direct cation-anion interactions.

Measurements in nonaqueous solvents indicated some concentration dependence of the chemical shifts particularly in acetonitrile solutions. Comparison of the magnesium-25 chemical shifts in nonaqueous solvents with the aqueous Mg^{2+} ion as reference showed that in all observed cases the shifts in nonaqueous solutions were downfield from the reference. The comparisons, however, were made for ~ 1.0 M solutions of magnesium salts.

This work is only in the initial stages and further studies should give a better indication of the behavior of magnesium-25 resonance in nonaqueous solutions.

SILVER-109 NMR

Silver-109 nmr measurements were made against saturated aqueous $AgClO_4$ solution as reference. Since the relaxation time of silver-109 nucleus is extremely long, the solutions were doped with small amounts of paramagnetic ions such as Ni^{2+}. In all cases the ratio of Ag^+ ion to that of the paramagnetic ion was at least 50/1. Unfortunately the sensitivity of silver-109 nmr is low (see Table I) and the lower limit of detection with our present instrumentation is 0.5 $\underline{M}$. Since the silver-109 chemical shift is strongly concentration dependent, it is not possible at this time to determine the infinite dilution shifts in different nonaqueous solvents.

Silver-109 chemical shifts were studied in water-acetonitrile mixtures as a function of solvent composition as shown in Figure 1. It is interesting to note that the plot is not monotonic but shows a definite and reproducible minimum at $\sim$ 0.6 mole fraction of CH_3CN. The most probable explanation of this singularity seems to be the existence of competing equilibria between the solvation and formation of contact ion pairs as the concentration of the organic solvent increases.

Preliminary results have been obtained on the study of silver complexes with aromatic compounds by silver-109 nmr. The addition of an aromatic compound (benzene, toluene, xylene) to a silver salt in a nonaqueous solvent results in a drastic downfield shift of the resonance as the concentration of the ligand is increased. Thus shifts of greater than 300 ppm were observed in THF solutions at ligand/Ag^+ mole ratios of 7:1.

STUDIES OF ALKALI ION COMPLEXES WITH MACROCYCLIC POLYETHERS AND

DIAZAPOLYOXABICYCLIC COMPOUNDS

For many years complexes of alkali metal ions were not considered to be an exciting area of research. Most chemists assumed that alkali complexes were neither stable nor important.

The advent of new macrocyclic ligands, such as polyethers (crowns) discovered by Pedersen (19) and of diazapolyoxabimacrocyclic ligands (cryptands) by Lehn and coworkers (20) opened a new era in the studies of alkali metal complexes. For the first time it became possible to obtain very stable complexes of these ions with stability constants comparable to those of transition metal complexes.

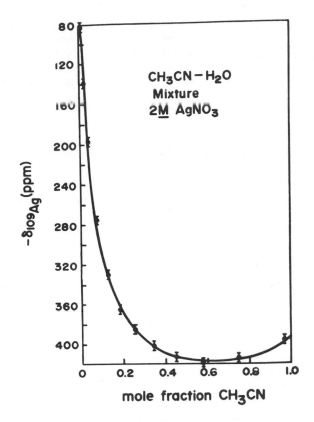

Figure 1.   Variation of the chemical shift of the silver-109
            resonance as a function of solvent composition for
            binary solvent water and acetonitrile. (Reproduced
            with the permission of Inorg. Nucl. Chem. Letters.)

        Since it is evident from the preceding discussion that the
alkali metal, magnesium-25 and silver-109 nmr measurements offer
a sensitive probe of the immediate environment of the respective
cations in solutions, it is obvious that the multinuclear nmr
technique can be very useful for the study of complexation reactions
and, in particular, for the study of crown and cryptate complexes
of alkali ions.

The IUPAC nomenclature for macrocyclic ligands is complicated and cumbersome. For the "crown" compounds, Pederson's suggested the use of trivial names which consist of (a) the number of the kind of substituent groups on the macrocyclic ring, (b) the total number of atoms in the ring, (c) the term "crown" followed by a number indicating the number of oxygen atoms in the ring. This suggestion has been essentially universally adopted by people working in this field. Thus the trivial name of compound I (Figure 2) is dicyclohexyl-18-crown-6.

Some limitations of the IUPAC nomenclature apply with even greater emphasis to the diazapolyoxamacrobicycles of Lehn et al. Lehn suggested that the ligands be called cryptands and their complexes cryptates. The number following the term indicates the number of oxygen atoms in each string. For example, compound II of Figure 2 is cryptand 211, usually abbreviated as C211, while compound III of the same figure is cryptand 222 (C222).

Several studies involving lithium-7, sodium-23 and cesium-133 nmr of crown and cryptate complexes have been reported in the literature.(21-26) In the case of crown compounds, the unsymmetrical electrical field gradient around an alkali nucleus results in a considerable broadening of the resonance lines. Only in the case of very narrow lines, such as those of cesium-133, can the chemical shifts be easily measured. Study of cesium-133 resonance of 18-crown-6-Cs$^+$ ion system in various solvents (25) showed large variations in the chemical shift as the ligand/Cs$^+$ mole ratio was changed (Figure 3). Particularly interesting behavior was observed in pyridine and, to a lesser extent, in acetone and propylene carbonate solutions where initial addition of the ligand to a cesium salt produced a downfield shift until a 1:1 mole ratio

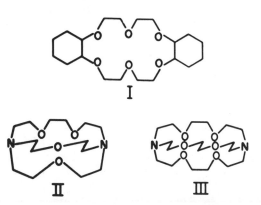

Figure 2.   I-Dicyclohexyl-18-Crown-6; II-Cryptand C211; III-
Cryptand C222.

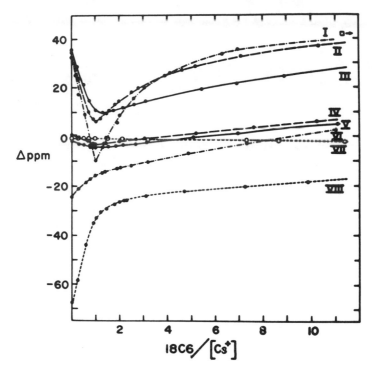

Figure 3.   Plot of cesium-133 chemical shift vs 18-crown-6/Cs$^+$ mole
ratio in different solvents. Concentration of cesium
salts is 0.01 M in all cases: (I) CsBPh$_4$ in PY, (II)
CsBPh$_4$ in Me$_2$CO, (III) CsBPh$_4$ in PC, (IV) CsBPh$_4$ in DMF,
(V) CsI in DMF, (VI) CsBPh$_4$ in MeCN, (VII) CsI in H$_2$O,
(VIII) CsBPh$_4$ in Me$_2$SO. (Reproduced with the permission
of the Journal of the American Chemical Society).

was reached, followed by an upfield shift as the concentration of
the ligand was further increased. This behavior is indicative of
a two-step reaction--initial formation of a stable (particularly
in pyridine) 1:1 complex, followed by the addition of a second
molecule of ligand to form a 2:1 "sandwich" complex.

It has been shown by Lehn (27) that the stability of cryptate
complexes is largely determined by the conformity in the size of
the three-dimensional cavity of the ligand and the ionic radius
of the de-solvated cation. Thus, in the case of lithium, the
most stable complex should be formed with cryptand C211 which has
a cavity size essentially identical to that of the Li$^+$ ion (1.6 Å
vs 1.56 Å). Cryptands C221 and C222 with cavity sizes of 2.2 Å
and 2.8 Å respectively, can form only relatively weak complexes
with Li$^+$. However, in this case, the effect of the solvent may
play a large role.

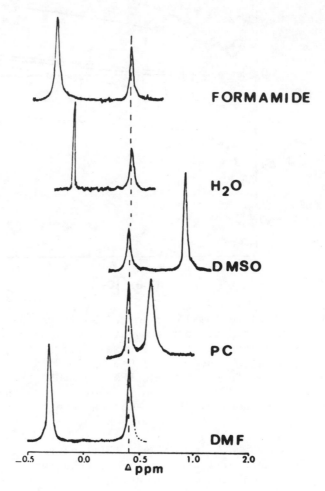

Figure 4.   Lithium-7 nmr spectra of lithium-C211 cryptate in
            various solvents: [C211] = 0.25 $\underline{M}$, [Li$^+$] = 0.50 $\underline{M}$.
            Chemical shift of Li-C211 is at 0.41 ppm $\underline{vs}$ aqueous
            LiClO$_4$ solution at infinite dilution.   (Reproduced with
            the permission of the Journal of Physical Chemistry.)

A study of lithium-7 resonance in Li$^+$-cryptand systems in various solvents clearly indicated the importance of the solvating ability of the solvents in the complexation reaction. With large cryptands, C222 and C221, there is little evidence for the lithium cryptate formation in dimethyl sulfoxide or in water. In a poor donor solvent, such as nitromethane, addition of C222 or C221 cryptand to a lithium perchlorate solution results in a drastic chemical shift which, however, becomes constant when the cryptand/Li$^+$ mole ratio reaches the value of 1:1. Thus the formation constant of the resulting cryptate must be fairly large. In a better donor solvent, pyridine, the limiting chemical shift is not reached even at 25:1 mole ratio due to the competitive action of the solvent.

On the other hand, with cryptand C211 the formation of a complex is not influenced by the donor ability of the solvent. In addition, the resonance frequency of the complexed cation is completely independent of the solvent (Figure 4). In this case, therefore, the ligand completely insulates the cation from the environment.

Moreover the lifetime of a cryptate complex is likewise influenced by the cavity/ionic radius relationship as well as by the solvent. In the case of Li$^+$-C222 and Li$^+$-C221 complexes, the exchange of Li$^+$ between the two sites (complex and free-solvated ion) is fast so that only one population-average resonance is observed. On the other hand, the exchange is much slower with the tight Li$^+$-C211 complex and, as seen in Figure 4. If excess of the metal ion is present, two signals are observed corresponding to the two sites of the Li$^+$ ion.

With the advent of Fourier transform nmr spectrometers and especially with the increasing availability (albeit, at a cost!) of high field superconducting solenoids, it is inevitable that the field of multinuclear nmr is expanding very rapidly. It is the hope of the author that the above brief discussion shows the great potential of this technique. So far, it has been applied to relatively few systems but it is obvious that it should have a bright future.

REFERENCES

1. C. W. Burges, R. Koschmieder, W. Sahm and A. Schwenk, Z. Naturforsch., 28a, 1753 (1973).

2. J. E. Wertz and D. Jardetzky, J. Chem. Phys., 25, 357 (1956).

3. D. Jardetzky and J. E. Wertz, J. Amer. Chem. Soc., 82, 318 (1960).

4. R. E. Richards and B. A. Yorke, Molecular Phys., 6, 289 (1963).

5. C. Deverell and R. E. Richards, ibid., 10, 551 (1966).

6. E. G. Bloor and R. G. Kidd, Can. J. Chem., 46, 3425 (1968).

7. R. H. Erlich, E. Roach and A. I. Popov, J. Amer. Chem. Soc., 92, 4989 (1970).

8. R. H. Erlich and A. I. Popov, ibid., 93, 5620 (1971).

9. V. Gutmann, "Coordination Chemistry in Nonaqueous Solvents," Springer-Verlag, Vienna, 1968.

10. M. S. Greenberg, D. M. Wied and A. I. Popov, Spectrochim. Acta, 29A, 1927 (1973).

11. G. E. Maciel, J. K. Hancock, L. F. Lafferty, P. A. Mueller and K. Musker, Inorg. Chem., 5, 554 (1966).

12. J. W. Akitt and A. J. Downs, Chem. Comm., 1966, 222.

13. Y. M. Cahen, R. P. Handy, E. T. Roach and A. I. Popov, J. Phys. Chem., 79, 80 (1975).

14. C. J. Jameson and H. S. Gutowsky, J. Chem. Phys., 40, 1714 (1964).

15. W. J. DeWitte, R. C. Schoening and A. I. Popov, Inorg. Nucl. Chem. Lett., 12, 251 (1976).

16. W. J. DeWitte, L. Liu, E. Mei, J. L. Dye and A. I. Popov, J. Solution Chem., In Press.

17. J. S. Shih and A. I. Popov, Inorg. Nucl. Chem. Lett., In Press.

18. L. Simeral and G. E. Maciel, J. Phys. Chem., 80, 552 (1976).

19. C. J. Pedersen, J. Amer. Chem. Soc., 89, 2495, 7017 (1967).

20. B. Dietrich, J. M. Lehn and J. P. Sauvage, Tetrahed. Letters, 1969, 2885.

21. J. M. Ceraso and J. L. Dye, J. Amer. Chem. Soc., 95, 4432 (1973).

22. J. P. Kintzinger and J.-M. Lehn, ibid., 96, 3313 (1974).

23. Y. M. Cahen, J. L. Dye and A. I. Popov, J. Phys. Chem., 79, 1289 (1975).

24. Y. M. Cahen, J. L. Dye and A. I. Popov, ibid., 79, 1292
     (1975).

25. E. Mei, J. L. Dye and A. I. Popov, J. Amer. Chem. Soc.,
     98, 1619 (1976).

26. A. M. Grotens, J. Smid and E. de Boer, Chem. Comm., 1971, 759.

27. J. M. Lehn, Structure and Bonding, 16, 1 (1973).

SODIUM-23 NMR STUDIES OF SODIUM CATION COORDINATION BY ORGANIC SOLVENTS

Christian Detellier and Pierre Laszlo

Institut de Chimie, Université de Liège

Sart-Tilman par 4000 Liège, Belgium

1. INTRODUCTION

Sodium-23 nuclear magnetic resonance is ripening rapidly as a tool of choice for study of sodium cation solvation. The ^{23}Na nucleus possesses highly favorable magnetic properties: 100% natural abundance, coupled with an intrinsic sensitivity 10^{-1} that of the ^{1}H nucleus; these properties make sodium 525 times more sensitive than the extensively-used ^{13}C nucleus.

Most of the exploratory work by Popov [1] involved the chemical shift; with a range of $\sim$ 40 ppm, it depends mainly on the nature of the sodium cation nearest neighbors. Exclusion of the counter ion from the sodium coordination shell, and its replacement by a solvent molecule (either through disaggregation or loosening of tight ion pairs), leads to a high-field shift [1,3]. If the solvent becomes a better donor, the sodium resonance moves downfield [2,4]. When two solvents compete for sodium coordination, the observed chemical shift as a function of solvent composition yields the relative complexing abilities; these have been determined by Greenberg and Popov [5] for all possible binary mixtures of nitromethane, acetonitrile, hexamethylphosphoramide, dimethylsulfoxide, pyridine, and tetramethylurea.

Sodium-23 ($I = 3/2$) is endowed with a quadrupole moment. The nmr linewidth arises from quadrupolar relaxation. We set upon a *joint* investigation of the sodium-23 chemical shift and linewidth, as independent parameters; any conclusion drawn simultaneously from both types of data would gain firmness. Furthermore, the linewidth depends upon the characteristics for reorientation of the sodium coordination shell, and provides additional information about solvation as a dynamic process.

We have studied the coordinating abilities towards sodium of oxygenated solvents (6,7), which bear structural resemblance to fragments from natural ionophores such as monensin or valinomycin. The sequence of complexing power we find closely matches that obtained by other spectroscopic or calorimetric methods. In addition to evaluation of thermodynamic parameters, the mechanism for solvent exchange between the bulk solvent and the first sodium solvation sphere is delineated (6,7).

2. EXPERIMENTAL

The ^{23}Na spectra were recorded in the Fourier transform mode, in order to achieve a better signal-to-noise ratio since the linewidths did not exceed 200 Hz, on a Bruker HFX-90 spectrometer. Chemical shifts δ are referred to an external reference of sodium chloride extrapolated to infinite dilution in water; they were corrected for the difference in magnetic susceptibility between sample and reference and were obtained to $\pm$ 0.2 ppm. In all cases, absorptions are true Lorentzians and are characterized by their width at half-height $\Delta\nu_{1/2}$. Viscosities of the solutions were measured using the apparatus of Desreux and Bischoff (8). All solvents used were carefully purified using standard methods (6). Sodium perchlorate was used at 0.1 M concentration, where it exists predominantly as loose, solvent-separated ion pairs (2). All quoted uncertainties are $\pm$ three standard deviations in order to take into account the systematic errors.

3. RESULTS

A. Pure Solvents

Sodium resonances in the pure solvents appear in Table 1. Two groups of solvents are evident from the chemical shifts; solvents coordinated to Na$^+$ via an ether oxygen, and characterized by δ values in the range 6-9.4 (tetrahydrofuran, and its 3-hydroxy derivative; the glymes), and solvents attached via an alcoholic OH oxygen, having δ values between 2.5 and 4.2 (methanol, ethylene glycol, tetrahydrofurfuryl alcohol, methyl cellosolve). Returning to the first group, of ether solvents, the glymes (MG, DG and TG) are characterized by very similar shifts, close to the value for THF. The intermediate value for 3-hydroxytetrahydrofuran is significant. Two interpretations are possible; coexistence of ion pairs in which sodium coordinates either to the THF oxygen or to the 3-hydroxy group, or a partial entry into the sodium coordination shell of the perchlorate counter-ions (as in a solvent-shared ion pair).

Table 1

Chemical Shifts and Reduced Linewidths for $Na^+ClO_4^-$ in
a Series of Oxygenated Solvents

Solvent (abbreviation)	$\delta (\pm 0.2$ ppm)	$\Delta\nu_{1/2}/\eta\ (\pm 0.1$ Hz.$mP^{-1})$
methanol (MeOH)	3.9	20.4^a
ethylene glycol (EG)	2.5	b
tetrahydrofurfuryl alcohol (THFu)	3.4	3.0_8
methyl cellosolve (MCS)	4.2	4.8_0
tetrahydrofuran (THF)	8.2	2.8_5
3-hydroxytetrahydrofuran (THF-3-OH)	6.0	3.1_9
monoglyme (MG)	8.6	4.6_7
diglyme (DG)	8.4	4.6_2
triglyme (TG)	9.4	4.8_9

———————

[a] bromide

[b] not determined

Turning to the reduced linewidths in Table 1, their non-zero
values have to be reconciled with an average symmetric environment
(T_d) of the sodium cation; analysis of the chemical shifts in
binary solvent mixtures (see below) points to tetracoordination by
monodentate ligands (MeOH, THF, THF-3-OH), while bidentate ligands
(the other solvents in Table 1) have two molecules in the sodium
coordination shell. Deviations from the T_d symmetry arise from
motions of the bound solvent molecules within the first coordination sphere; non-zero electric field gradients thus set at the
sodium nucleus will fluctuate in direction and in amplitude at the
rate of these internal reorientations. These occur at a slightly
lower rate than reorientation of bulk solvent molecules. The
orientational correlation time for water molecules *bound* to Na^+
is less than twice that for the *free* water molecules, with an
activation energy for exchange below 400 cal/mole (9).

A similar conclusion is reached in a phenomenological approach of the Debye-Stokes-Einstein vein; a Stokes radius of 3.6 ± 0.7 Å is obtained (10) in accord with the conductimetric measurements by Szwarc et al. (11) of 3.95 ± 0.5 Å (THF) and 3.2 ± 0.4 Å (MG). The corresponding nmr Stokes volume of 120 ± 20 ml/mole is clearly much too small for a rigid assembly of four THF molecules or two MG molecules coordinated to the sodium cation. The statement: "the lone pairs of the oxygen atoms of the ether molecules are shared by the Na^+ ion to form a coordination complex with a very rigid structure, so that the ion with its first solvation sphere may be considered as one chemical entity" (12) is not applicable *sensu stricto* to our linewidth data.

These linewidths are consistent with *fast* exchange of solvent molecules between the first solvation sphere and the bulk of the solution. The corresponding 120 ml Stokes volume is that of a spherical particle centered on sodium, having a radius of ~ 3.75 Å. This value can be regarded as the ~ 2.35 Å sodium-oxygen distance (13,15) augmented by the van der Waals radius for oxygen of 1.40 Å. The sodium solvation shell is kinetically labile; exchange proceeds at a rate commensurate with that of the overall reorientational motion, while maintaining the oxygen heteroatoms at van der Waals contact with the sodium cation.

B.   Binary Solvent Mixtures

The results obtained with the THF-THFu mixture are taken as example for treatment of the data (Table 2). In order to analyze the monotonous dependence of the chemical shifts upon solvent composition, we have tested the following models:

4 THF   →   4 THFu

4 THF   →   3 THFu (with a dissociative second step)

4 THF   →   3 THFu (with a dissociative first step)

4 THF   →   2 THFu

For each of six experimental points j, we calculate from the Covington equations (16) the average value of the equilibrium constants for each combination *including* this point (Tables 3-6).

The following succession of equilibria adequately describes the results:

$$(Na^+)_{4 \text{ THF}} \overset{K_1}{\underset{}{\rightleftarrows}} (Na^+)_{2 \text{ THF, THFu}} \overset{K_2}{\underset{}{\rightleftarrows}} (Na^+)_{2 \text{ THFu}}$$

Table 2

Sodium-23 Chemical Shifts in Binary Mixtures of
Tetrahydrofuran and Tetrahydrofurfuryl Alcohol

[THFu], mole fraction	δ(ppm)
0.000	8.17 ± 0.08
0.019	6.85 ± 0.08
0.039	6.37 ± 0.13
0.078	5.83 ± 0.08
0.177	5.21 ± 0.08
0.317	4.52 ± 0.13
0.428	4.24 ± 0.13
0.584	4.01 ± 0.25
0.705	3.69 ± 0.13
0.796	3.49 ± 0.21
0.920	3.72 ± 0.13
1.000	3.43 ± 0.17

Table 3

Test of the (Invalid) 4 THF → 4 THFu Model

	K_1	K_2	K_3	K_4
$\bar{1}$	-99	28	4	2
$\bar{2}$	88	34	6	2
$\bar{3}$	115	38	8	2
$\bar{4}$	-176	36	8	2
$\bar{5}$	-130	19	10	2
$\bar{6}$	-130	26	11	1.4

Table 4

Test of the (Invalid) 4 THF → 3 THFu Model, In Which the
Dissociative Step is the Second

	K_1	K_2	K_3	K_4
$\bar{1}$	-881	-0.3	- 11	7
$\bar{2}$	-520	-11	- 10	18
$\bar{3}$	-1356	-15	- 28	17
$\bar{4}$	-1210	-45	6	20
$\bar{5}$	-920	-30	10	18
$\bar{6}$	-1450	-50	9	25

Table 5

Test of the (Invalid) 4 THF → 3 THFu Model, In
Which the Dissociative Step is the First

	K_1	K_2	K_3	K_4
$\bar{1}$	-184	615	14.6	1.9
$\bar{2}$	-147	617	8.0	1.8
$\bar{3}$	-200	634	6.6	2.0
$\bar{4}$	-216	510	2.3	2.0
$\bar{5}$	-190	466	6.3	2.0
$\bar{6}$	-102	340	5.1	1.8

Table 6

The Successful 4 THF → 2 THFu Model

	K_1	K_2
$\bar{1}$	712	19
$\bar{2}$	682	20
$\bar{3}$	760	21
$\bar{4}$	892	20
$\bar{5}$	484	20
$\bar{6}$	782	17
$\bar{1}\bar{2}\bar{3}\bar{4}\bar{5}\bar{6}$	719 $\pm$ 136	20 $\pm$ 1

The tetracoordination reported here

(i)   has precedence; Day and coworkers have reported a coordination number of 4 THF molecules per sodium cation (17);

(ii) is not surprising for organic solvents with molecular volumes significantly larger than that of water (CN = 6) (18);

(iii) is taken to mean strictly the number of nearest neighbor solvent molecules, as "seen" by ^{23}Na nmr, whose chemical shifts are a local property, little sensitive to more distant perturbations. The calculations by Clementi et al. (14) show equidistant water molecules for the $(\overline{Na^+})_2H_2O$, $(Na^+)_3H_2O$, and $(Na^+)_4H_2O$ clusters, whereas $(Na^{\mp})_6H_2O$ is a distorted octahedron, with two water molecules at a somewhat greater distance from the central ion.

It is impossible to measure directly the sodium-23 chemical shift characteristic of the intermediate structure $(Na^+)_2$ THF, THFu. For none of the binary solvent mixtures we have studied (6), the chemical shift wanders outside the range between the limiting pure solvents at mole fractions of zero and unity. We have made the usual assumption (5,16) of the arithmetic mean for the chemical shift of all these intermediates; in this example, δ(THF,THFu) = $\frac{1}{2}[\delta$THF + δ(THFu)]. In the present work, this assumption appears to be sustained by the resulting equilibrium constant ratios, close to the statistical values (6), and by the detailed analysis of the changes in the linewidths with composition of the binary solvent mixtures (7). The resulting Gibbs free enthalpies ΔG_{tr} for transfer between a reference solvent-arbitrarily chosen here as THF - and a given solvent are indicated in Table 7.

4. DISCUSSION

For six solvents common to both studies, the sequence of ΔG_{obsd}° (Table 7) is almost superimposable with that found by Arnett et al. (19) for the ΔH_{obsd} in complexation of Na^+ by hydroxy and ether solvents, in acetone solution:

$$THF < CH_3OH < G < DG < EG < TG \qquad \text{(Arnett)}$$

$$THF < CH_3OH < G < DG < TG < EG \qquad \text{(this study)}.$$

Our results for the glymes (MG, DG, TG) are approximately linear with the number of oxygen atoms in these molecules (2,3,

Table 7

ΔG° Values for Transfer of $Na^+ClO_4^-$ Between
THF and the Listed Solvents

Solvent	G_{tr} ($\pm$ 0.1 kcal/mole)
THF	0.0
MG	- 3.7
DG	- 5.3
TG	- 6.4
MeOH	- 1.6
EG	- 6.6
THFu	- 5.7
THF-3-OH	- 3.4

and 4 respectively), with an increment $\sim$ -1.3 kcal/mole per additional oxygen.

Protic solvents can stabilize the anion by hydrogen bonding;
thus, methanol is "better" than tetrahydrofuran by 1.6 kcal/mole.
One would then predict an additional stabilization of 2 x 1.6 :
3.2 kcal/mole for ethylene glycol when compared to dimethoxyethane
(MG); the observed difference, 2.9 kcal/mole, comes close. By the
same token, since tetrahydrofurfuryl alcohol (THFu) and dimethoxyethane (MG) are bidentate ligands having in common a $-OCH_2CH_2O-$
fragment, the former should also be stabilized with respect to the
latter by ca. 1.6 kcal/mole; the observed value is 2 kcal/mole.
Interestingly, THF-3-OH, in which the OH group is also two carbons
removed from the oxygen donor atom, benefits from a stabilization
greater by a factor 2 (3.4 kcal/mole, when compared to THF).   In
this case, we believe that shared ion-pairs are present; hydrogen-
bonding by the 3-OH function maintains the perchlorate anion in the
vicinity of the Na$^+$ cation. This explanation is also consistent
with the evidence from the chemical shifts in the pure solvents
discussed above in Section 3.

Consider again the results for the glymes: why does the
complexing aptitude towards sodium increase in the sequence
G < DG < TG, as observed by many authors (19-23)? A crucial finding is that by Arnett (19): the entropy change ΔS° upon complexation
of Na$^+$ by *either* monoglyme or tetraglyme, in acetone solution, is
almost identical (- 12 e.u.)! The negative value is not consistent
with a significant gain in translational entropy, upon entry of
one glyme molecule in the Na$^+$ solvation sphere to replace *two*
acetone molecules. The fact that ΔS° is the same for *mono*glyme and
*tetra*glyme implies identical losses in the entropy of internal
rotation for both ligands, as they wrap themselves around the
cation (provided the stoichiometry remains the same). It has just
been shown that entropy loss in the gas phase is strikingly similar
for one bidentate ligand as for similar monodentate ligands (24),
as if the losses of rotational degrees of freedom incurred during
the quasi-cyclization of the bidentate ligand did not show in the
net entropy change.

The same authors (24) have shown that the "chelate effect"
does not operate in the gas phase: two monodentate ligands give
an enthalpy lowering greater by $\sim$ 40% to that produced by a single
bidentate ligand; this is more than sufficient to compensate for
the unfavorable translational entropy loss for one more monodentate
ligand. The "chelate effect" found in solution must originate in
lower dipolar repulsions, and a weaker polarization of the bidentate
ligand upon single point attachment (24).

Thus, one should refrain from differentiating the glyme
molecules from one another on the basis of internal rotation

degrees of freedom (6); the values for $\Delta\nu_{1/2}/_\eta$ (Table 1) are also very similar for the glymes (MG, DG, TG): and do not support a higher structuration of the DG and TG solvation sphere as compared to MG (6). If the entropy change upon transfer from THF to a glyme solvent is invariant upon the number of coordination sites in the glyme molecule, then the glymes rank themselves in the order MG < DG < TG on the basis of the enthalpy term predominantly.

Introduction of an oxygen atom into the Na^+ solvation sphere suffers from the electrostatic dipolar repulsion from the oxygen atoms already present there. For instance, the $-\Delta H°$ term is 24.0 kcal/mole for the first bound water molecule, 19.8 for the second, 15.8 for the third, 13.8 for the fourth, and so forth (25). Likewise, when THF interacts with Na^+ in cyclohexane solution, the $-\Delta G°$ values (statistically corrected) we infer from the results by Day et al. (17) are 1.76 kcal/mole for the second THF molecule, 1.16 for the third, and 0.77 for the fourth. When two (or more) oxygen atoms belong to the same molecule, they are maintained together by covalent bonding in spite of their electrostatic repulsion; i.e. upon cation complexation, one does not have to draw any longer upon the Coulombian ion-dipole attraction in order to pay the price for the dipole-dipole repulsion. This price has already been paid during synthesis of the polydentate ligand. This analysis, in terms of a predominant enthalpy factor, has already been presented by others for the macrocyclic effect (26).

5.   CONCLUSIONS

Gathering the various elements at hand, one can attempt to draw a mental picture of the sodium cation solvation sphere. In the time scale characteristic of nuclear magnetic resonance, solvent molecules are drawn towards the cation in a tetracoordinate arrangement, by the ion-dipole attraction. They mutually repel one another, both sterically and by the dipole-dipole repulsion. These forces lead to a definite structuring around the Na^+ cation. However, the details of the interaction are relatively unimportant. Two examples follow. In the molecule of tetrahydrofurfuryl alcohol, the non-equivalent THF ether oxygen and the OH oxygen in the side chain have interactions of comparable magnitude with the sodium cation. Conformations can be drawn for diglyme and triglyme to make each of their three or four oxygen atoms interact simultaneously with the sodium cation, in a manner highly reminiscent to that of crown ethers (6,15). Yet, the price to be paid for freezing supplementary internal rotation degrees of freedom with respect to glyme is prohibitive: approximately 2.7 e.u. per C-O or C-C bond frozen into one of its three rotamers amounts to a total of $\sim$ 4 kcal/mole disfavoring DG with respect to G, or TG with respect to DG, in such highly ordered conformations. The system prefers a compromise in which, even though MG is not an ideal bidentate ligand (with standard bond lengths and angles, and a 2.55 Å sodium-oxygen distance, the O-Na-O angle is 72°, i.e. much smaller than the tetrahedral value), DG and TG bind sodium in a similar manner as MG; probably no more than two oxygens bind sodium simultaneously in DG and TG.

These solvation spheres have dynamic structures constantly forming and breaking. Solvent molecules exchange with the sur-rounding bulk solvent, with small energy barriers. A given solvent molecule has a residence time, in the solvation shell, of the same order of magnitude as the orientational correlation time for bulk solvent molecules. For such a univalent cation as Na^+, the solvation enthalpies are not strong enough that solvent molecules coordinate the cation for a time long with respect to the correla-tion times, so that a true complex is formed. The macroscopic analog of this statement is that organic solvent molecules incur a negligible loss in translational entropy from their interaction with the sodium cation. Thus, we are provided with an interesting case of positive solvation: weaker than a true complex formation, but stronger than a transient molecular interaction with the solvent molecules, of the collision complex type. Relaxation studies of water and methanol solutions have led to a similar conclusion concerning solvation of the sodium cation (27); a fully-oriented shell of water (or methanol) solvent molecules is present around the sodium cation, with the solvent dipoles preferentially oriented towards it.

6.   SUMMARY

Sodium-23 nuclear magnetic resonance is useful in the exami-nation of solvation of the sodium cation by oxygen containing molecules. Four monodentate ligands or two bidentate ligands surround the Na^+ ion. Free energies are obtained for transfer of sodium perchlorate between various solvents in which it exists as loose ion pairs. Their values show the usual sequence of complex-ing power in the glyme series, with additional stabilization for hydrogen bonding of the perchlorate anion in protic solvents.

7.   ACKNOWLEDGMENTS

We thank Fonds de la Recherche Fondamentale Collective for a grant in aid of the purchase of the nmr spectrometer, and IRSIA for award of a pre-doctoral fellowship to C.D. We gratefully acknowledge Professeur D'Or and the Académie Royale des Sciences (Fondation Agathon de Potter) for a travel grant to the meeting in San Francisco for P.L.

8.   REFERENCES

(1)   A. I. Popov, Pure Appl. Chem., 41, 275 (1975).

(2)   C. Detellier and P. Laszlo, Bull. Soc. Chim. Belges, 84, 1081 (1975).

(3)   E. G. Bloor and R. G. Kidd, Can. J. Chem., 46, 3425 (1968).

(4)   R. H. Erlich, E. Roach, and A. I. Popov, J. Am. Chem. Soc., 92, 4989 (1970); R. H. Erlich and A. I. Popov, J. Am. Chem. Soc., 93, 5640 (1971).

(5)   R. H. Erlich, M. S. Greenberg, and A. I. Popov, Spectrochim. Acta, 29A, 543 (1973); M. S. Greenberg and A. I. Popov, Spectrochim. Acta, 31A, 697 (1975).

(6)   C. Detellier and P. Laszlo, Helv. Chim. Acta, 59, 1346 (1976).

(7)   C. Detellier and P. Laszlo, Helv. Chim. Acta, 59, 1346 (1976).

(8)   V. Desreux and J. Bischoff, Bull. Soc. Chim. Belges, 59, 93 (1950).

(9)   B. P. Fabricand, S. S. Goldberg, R. Leifer, and S. G. Ungar, Mol. Phys., 7, 425 (1964); G. Engel and H. G. Hertz, Ber. Bunsenges, Phys. Chem., 72, 808 (1968).

(10)   C. Detellier and P. Laszlo, Bull. Soc. Chim. Belges, 84, 1087 (1975).

(11)   C. Carvajal, K. J. Tölle, J. Smid, and M. Szwarc, J. Am. Chem. Soc., 87, 5548 (1965).

(12)   G. J. Hoytink, Chem. Phys. Letters, 31, 21 (1975).

(13)   P. Schuster, W. Jakubetz, and W. Marius, in Topics in Current Chemistry, 60, 1 (1975).

(14)   H. Kistenmacher, H. Popkie, and E. Clementi, J. Chem. Phys., 61, 799 (1974).

(15)   M. Dobler, J. D. Dunitz, and P. Seiler, Acta Cryst., B30, 2741 (1974).

(16)   A. K. Covington, T. H. Lilley, K. E. Newman, and G. A. Porthouse, J. Chem. Soc. Faraday I, 69, 963 (1973); A. K. Covington, K. E. Newman, and T. H. Lilley, J. Chem. Soc. Faraday I, 69, 973 (1973); A. K. Covington, I. R. Lantzke,

and J. M. Thain, J. Chem. Soc. Faraday Trans. 1., 1869 (1974); A. K. Covington and J. M. Thain, J. Chem. Soc. Faraday Trans. I., 1879 (1974).

(17) E. G. Höhn, J. A. Olander, and M. C. Day, J. Phys. Chem., 73, 3880 (1969).

(18) A. T. Tsatsas, R. W. Stearns, and W. M. Risey, J. Am. Chem. Soc., 94, 5247 (1972).

(19) E. M. Arnett, H. Chung Ko, and C. C. Chao, J. Am. Chem. Soc., 94, 4776 (1972).

(20) J. F. Garst in "Solute-Solvent Interactions", J. F. Coetzee and C. D. Ritchie, Ed., Marcel Dekker, New York, N.Y., 1969.

(21) H. E. Zaugg, J. Am. Chem. Soc., 83, 837 (1961).

(22) J. L. Down, J. Lewis, B. Moore, and G. Wilkinson, J. Chem. Soc., 3767 (1959).

(23) J. Smid in "Ions and Ion Pairs in Organic Reactions", vol. 1, M. Szwarc, Ed., chapter 3, Wiley-Interscience, New York, N.Y., 1972.

(24) W. R. Davidson and P. Kebarle, Can. J. Chem., 54, 2594 (1976).

(25) I. Dzidic and P. Kebarle, J. Phys. Chem., 74, 1475 (1970).

(26) inter alia: P. Paoletti, L. Fabrizzi, and R. Barbucci, Inorg. Chem., 12, 1961 (1973); Inorg. Chim. Acta Rev., 7, 43 (1973); A. Dei and R. Gori, Inorg. Chim. Acta, 14, 157 (1975); E. Kauffmann, J. M. Lehn, and J. P. Sauvage, Helv. Chim. Acta, 59, 1099 (1976).

(27) C. A. Melendres and H. G. Hertz, J. Chem. Phys., 61, 4156 (1974).

APPLICATION OF SPECTROSCOPIC AND ELECTROCHEMICAL TECHNIQUES IN

STUDIES OF CHEMISTRY OF RADICAL ANIONS AND DIANIONS

M. Szwarc

Department of Chemistry, State University of New York
College of Environmental Science and Forestry
Syracuse, New York   13210

In this chapter I intend to review some of the approaches
that utilize electrochemical and spectroscopic techniques to
provide information about the nature of ionic aggregates, such as
ion pairs, triple ions, etc. I shall also show how these
approaches can be employed in studies of labile and short-lived
species.

Ion pairs and higher ionic aggregates often exist in a
variety of forms, e.g., tight ion pairs are distinct from loose
ones. Spectroscopic techniques, such as NMR, infrared and Raman
spectroscopy, visible and UV spectroscopy, and especially ESR
reveal their existence and allow us to deduce their structure.
ESR spectroscopy is limited; it can be used only in studies of
paramagnetic species. Nevertheless, in spite of this limitation
its power is astonishing--it may reveal the fine details of
structure of ionic aggregates and the dynamics of their behavior.
The literature pertaining to that subject is too voluminous to be
discussed here, and the interested reader may be referred to recent reviews (1) dealing with some vigorously investigated topics.
However, a few examples illustrating the use of spectroscopic and
electrochemical techniques in studies of these problems will be
outlined here with emphasis on the ways by which the conclusions are
drawn from the results.

The pioneering work of Weissman (2) provided the first positive evidence for the existence of ion pairs. He reported that
the 25-line ESR spectrum of a free naphthalenide ion, shown in
Fig. 1, changes into a 100-line spectrum when the radical anion
becomes associated with a diamagnetic sodium cation. An example

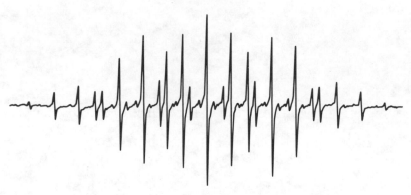

Figure 1.   The ESR spectrum of the free naphthalenide anion in HMPA.

of the latter spectrum is seen in Figure 2; it refers to sodium naphthalenide dissolved in tetrahydropyran (THP). Each of the 25 lines characterizing the free naphthalenide anion is split into 4 equidistant lines, the splitting resulting from the interaction of the odd electron's spin of the anion with the 3/2 spin of the sodium nucleus. The strength of this interaction is reflected in the magnitude of the coupling constant $\underline{a}$ measured by the separation of the equidistant new lines; in the above example $\underline{a}$ = 1.4 G. The splitting of the ESR lines not only demonstrates the occurrence of an anion-cation association but also proves that the associate is relatively long lived (> 10^{-5} sec), because the split lines are sharp.

Addition of small amounts of tetraglyme ($\sim 10^{-3}$ M) to $\sim 10^{-4}$ M solution of sodium naphthalenide in THP drastically changes the ESR spectrum (3). This is seen on comparing Figures 2 and 3. The coupling to sodium nucleus is reduced from 1.4 to 0.4 G implying that the addition of the glyme leads to conversion of the relatively tight pairs originally present in THP into looser ones. Apparently tetraglyme strongly solvates the Na^+ cations and thus separates them from the naphthalenide anions. In fact, judicious choice of the glyme concentration allowed us to record ESR spectra simultaneously revealing the lines of the tight and loose pairs. From their relative intensities and width we deduced the equilibrium and rate constants of the reaction,

tight sodium naphthalenide pairs + tetraglyme $\rightleftarrows$ loose sodium naphthalenide pairs.

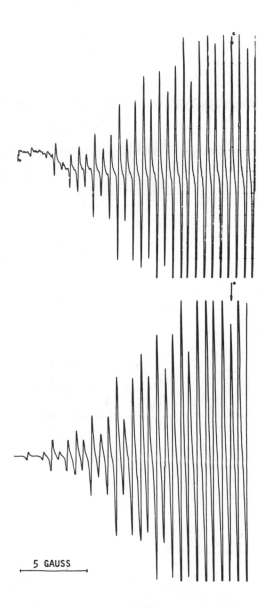

Figure 2. The ESR spectrum of sodium naphthalenide pairs in THP (below the computer simulated spectrum). The coupling constant to sodium is 1.4 G.

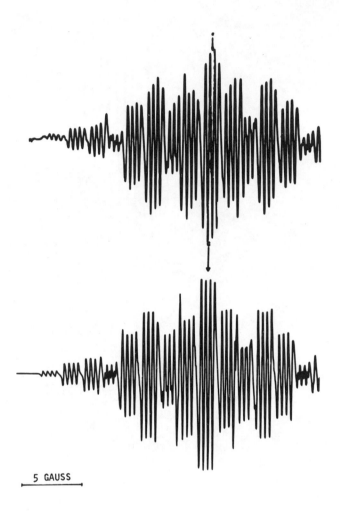

5 GAUSS

Figure 3. The ESR spectrum of $\sim 10^{-4}$ M solution of sodium naphthalenide in THP after addition of tetraglyme (its concentration $\sim 10^{-3}$ M) (below the computer simulated spectrum). The coupling constant to sodium is 0.4 G.

Extensive ESR studies of equilibria and kinetics of interconversion of tight into loose pairs in pure solvents were described by Hirota and his associates (4) and similar investigations were reported later by other workers.

Studies of Gough (5) take us one step further. His ESR investigations revealed the formation of triple ions, e.g., Na$^+$, duroquinone$\bar{\cdot}$, Na$^+$, the presence of which is evident from the splitting of each of the 13 lines of the free duroquinone radical anions into seven lines, each resulting from the odd electron spin interaction with two Na nuclei having total spin of 3.

The power of ESR spectroscopy is further manifested by the phenomenon of line width alteration. Its significance is illustrated by the ESR spectrum of alkali salt of quinone radical anion. The spectrum of the free quinone radical anion consists of 5 equidistant lines arising from the interaction of the odd electron spin with 4 equivalent protons, their intensity being 1:4:6:4:1. The association of this anion with an alkali cation places the latter in the vicinity of one or the other oxygen center because they bear most of the negative charge. Such an asymmetric placement of the cation destroys the equivalence of the 4 protons; however, for those ions having all protons of the same spin, i.e., of total spin 2 or -2, the energy of the system remains unaltered when the cation hops from one end to the other. Hence, the wing ESR lines are sharp and unaffected by the hopping. This is not so for the second and fourth lines corresponding to total spin 1 or -1. In those species one 0 center is flanked by two protons of the same spin whereas the other finds two protons of opposite spin in its neighborhood. For these species the energy of the system is modulated by the cation hopping and this broadens the respective lines. Finally, the center line represents those spin-populations where two protons have spin 1/2 while the other two -1/2, i.e., the species of total spin 0. Four of them are symmetric in relation to proton hopping, namely,

and their energy is not affected by the hopping, whereas the energy of the remaining two kinds, namely,

is modulated. Thus the center line remains sharp but its intensity
is reduced from 6 to 4. Analysis of such a spectrum permits one
to find the rate of intramolecular cation hopping from one center
of association to the other.

Let us pass now to the effect of ion-pairings and of their
state of solvation on their UV and visible spectra. The energy
of the ground state of an ion decreases when it becomes associated
with a counterion. However, the energy of an excited state is less
affected by the pairing. The counterion in non-excited associate
is located in a place most favorable for the anion-cation interaction. The optical excitation perturbs the distribution of the
negative charge and reduces the extent of this interaction because,
due to Frank-Condon restriction, the counterion cannot move to the
new most favorable location. Thus the pairing leads to a hypso-
chromic shift of the absorbance.

Conversion of a loose pair into a tight one affects the
absorption spectrum in a way similar to that discussed above,
namely the absorption of a tight pair is shifted toward shorter
wavelength when compared with a loose one. This phenomenon was
first described by Hogen-Esch and Smid (6) who studied the optical
spectra of salts of fluorenyl carbanions (Fl^-). The absorption of
sodium fluorenyl in tetrahydrofuran (THF) shows two maxima at 355
and 373 nm. Their relative intensities are temperature dependent,
the 373 nm peak grows while the 355 nm one wanes as the temperature
of the solution is lowered. It was shown (6) that the two peaks
represent tight and loose Fl^-, Na^+ ion pairs coexisting in equilibrium,

$$\text{tight } Fl^-, Na^+ \rightleftarrows \text{loose } F^-, Na^+.$$

The conversion of the tight pair into the loose one is exothermic;
the increasing solvation of the sodium cation by THF more than
compensates for the loss of Coulombic energy resulting from the
partial separation of the oppositely charged ions. Hence, the
equilibrium shifts to the right at lower temperatures leading to
the observed spectral changes from which ΔH and ΔS of the conversion was calculated.

The conversion of tight pairs into loose ones is associated
with decrease of total volume of the solution due to electrostric-
tion. Therefore, increasing hydrostatic pressure shifts the
equilibrium of their interconversion toward the loose pairs, again
leading to the appropriate changes in their spectrum. These
spectral changes led to the determination of ΔV of the conversion
(7), and from its value one can deduce the magnitude of the forces
squeezing the solvent molecules together around the cation.

The spectral studies of the tight-to-loose pairs equilibria led to most interesting and quantitative information about the strength of interaction of solvent with alkali and alkali-earth cations. This subject has been reviewed recently (1) but one example of such studies, showing an unusual behavior of these systems, will be considered here.

Addition of crown ethers to solution of tight fluorenyl pairs converts them into loose pairs (1), a phenomenon again revealed by the bathochromic shift of their absorbance. Barium fluorenyl (Ba^{2+},Fl_2^-) in THF absorbs sharply at 347 nm and the position of that peak implies that under these conditions the fluorenyl anions are tightly bound to Ba^{2+} cation. It could be expected that the addition of crown ethers should transform this aggregate into a loose one. Indeed, the absorption spectrum was changed; two peaks were observed (8), one at 347 nm, the other at 372 nm, and from their intensities it was deduced that a 50:50 mixture of tight and loose species was produced. Their proportion was not affected by temperature, even at -70°C. It was concluded, therefore, that crown ether is placed asymmetrically forming a $Fl^-,Ba^{2+},Crown,Fl^-$ complex. However, its NMR spectrum does not reveal the asymmetry indicating that the Ba^{2+} oscillates through the ether's cavity making both Fl^- identical on the NMR time scale.

A striking example of complexity of some ionic aggregates was provided by the lithium salt of tetraphenylethylene dianion (9) $(T^{2-},2Li^+)$. The absorbance of the free dianion and its sodium salt shows only one broad band centered around 485 nm. The absorption of the lithium salt is strikingly different as shown by Figure 4; the molar absorbance of the 485 peak decreased and a new strong absorption band appeared at 390 nm. It seems that the intensity of the new band was derived from 1/2 of the intensity of the 485 nm band appearing in the spectrum of the free dianion. These spectral changes permit the titration of $T^{2-},2Na^+$ by LiCl. Gradual additions of the latter salt to THF solution of the former result in a decrease of intensity of the 485 band and growth of the 390 nm peak, simultaneously with the precipitation of sodium chloride.

The shape of the absorption spectrum of $T^{2-},2Li^+$ is not affected by temperature, but the new 390 nm band wanes and the 485 nm grows on addition of tetraglyme. Studies of this phenomenon indicate that one molecule of tetraglyme suffices to convert the $T^{2-},2Li^+$ absorbing at 390 and 485 nm into a "glymated" $T^{2-},2Li^+$ absorbing at 485 nm only. Obviously in the absence of the glyme the two Li^+ cations are different, one being strongly solvated by THF forming a loose associate with the $-\bar{C}Ph_2$ chromophore of $Ph_2\bar{C}.\bar{C}Ph_2$ (T^{2-}), while the other Li^+ is tightly bound to the other part of T^{2-}. The observed spectral changes are not caused

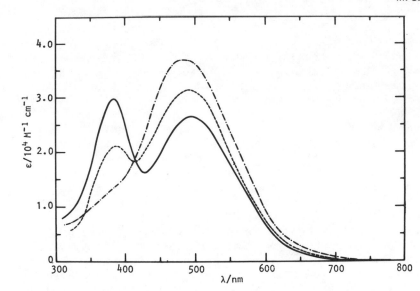

Figure 4.   Absorption spectra of $T^{2-},2Na^+$ (dot-dash line),
$T^{2-},2Li^+$ (solid line), and 50-50 mixture (dashed line).

by a chromophore-chromophore splitting and the ^{13}C NMR demonstrates
that the cations exchange their roles in the aggregate, figura-
tively

$$PH_2\bar{C} - \bar{C}pH_2 \rightleftharpoons Ph_2\bar{C} - \bar{C}pH_2 \ .$$

Even more striking is the spectrum of Ba^{2+} salt of tetraphenylethylene dianion (10). This salt, like the lithium salt,
has two absorption bands, the "conventional" one at 485 nm and the
"new" one at 350 nm, both of the same intensity. Apparently the
Ba^{2+} cation is placed asymmetrically, being very strongly bonded
to one $-\bar{C}Ph_2$ chromophore, leaving the other as a psuedo-free anion.
The great strength of the bonding is responsible for the astonish-
ingly large shift of the absorbance by 130 nm.

The following approach is useful in studies of solvation of
some radical anions (11). Aromatic hydrocarbons may be reduced by
alkali metals to the respective radical anions. The reaction,
alkali metal + A (solution) $\overset{K^o}{\rightleftharpoons}$ $A^{\bar{\cdot}},Alk^+$ (solution) where A denotes
an aromatic hydrocarbon, leads to an equilibrium mixture with the
ratio $[A^{\bar{\cdot}},Alk^+]/[A] = K_0$ independent of the concentration of A but
determined by the nature of the hydrocarbon, of the solvent, kind
of the alkali metal, and by temperature. By a judicious choice of

A and solvent this ratio may be kept small, e.g., for biphenyl in
tetrahydropyran (THP) or naphthalene in diethylether reduced by
sodium metal. Addition of solvating agent affects this ratio
because the reactions with the solvating agent $S\ell$, such as

$$A^{\bar{\cdot}},Alk^{+} + S\ell \overset{K_1}{\rightleftarrows} A^{\bar{\cdot}},Alk^{+},S\ell$$

$$A^{\bar{\cdot}},Alk^{+},S\ell + S\ell \overset{K_2}{\rightleftarrows} A^{\bar{\cdot}},Alk^{+}S\ell_{2}, \text{ etc.,}$$

convert the original pairs into solvated ones and thus more of A
has to be reduced to $A^{\bar{\cdot}},Alk^{+}$ to maintain the primary equilibrium
governed by K_o.

The apparatus depicted in Figure 5 facilitates studies of such
systems. A round-bottomed flask with an alkali-metal mirror
deposited on its walls is half-filled with a solution of the
investigated hydrocarbon A. An all-glass centrifugal pump circu-
lates the solution from the flask, through a flat optical cell
placed in the compartment of a spectrophotometer, back into the
flask. The temperature of the solution is controlled by keeping
the flask in a thermostated bath.

The reaction of the dissolved hydrocarbon A with the mirror
maintains the equilibrium concentration of $A^{\bar{\cdot}},Alk^{+}$, the concentration being calculated from the directly measured absorbance of the
optical cell. Thus K_o is determined for a series of temperatures.
Thereafter a desired amount of a solvating agent is added to the
solution and the absorbance redetermined at each of the studied
temperatures. This procedure is then repeated with increasing
amounts of solvating agents. Since a new kind of solvated pairs
is formed from the original ones, the latter have to be re-
plenished by the reaction of A with the mirror.

The shape of the absorption spectrum is not altered when the
molecules of the solvating agent do not separate the associated
ions but only replace some of the molecules of the original solvent that surrounded the ion pairs. This merely increases its
intensity. In such a case the increase of the absorbance Δa
arising from the addition of the solvating agent is given by
$\Delta a/a_o = K_1 C_s + K_1 K_2 C_s^{2}...$, where C_s is the concentration of the
solvating agent, provided that its activity coefficient is unity.
For dilute solutions this is a plausible approximation. However,
when the solvating agent separates the associated ions, two peaks
appear in the absorption spectrum, and from their intensities
their relative concentrations can be evaluated. Thus, this
procedure permits us to distinguish between the external and
internal solvation since the former increases the intensity of the
spectrum without producing any new absorption peak while the latter

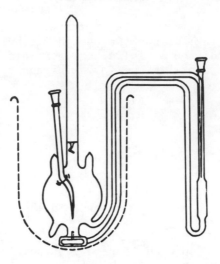

Figure 5. Schematic drawing of the apparatus used in studies of solvation of ion pairs produced by reduction of aromatic hydrocarbons.

results in the appearance of a new absorption band. The spectro-photometric results were supplemented by ESR studies.

This approach revealed the first example (11a) of a system in which the externally solvated ion pairs are in equilibrium with an internally solvated pair. Such an equilibrium is described by the equation $A^{\bar{\cdot}}, Alk^+, S\ell \rightleftarrows A^{\bar{\cdot}}, S\ell, Alk^+$ and the results allowed us to determine the respective K, ΔH and ΔS.

Much information about the nature of ion pairs solvation have been deduced from studies of disproportionation of radical anions $(A^{\bar{\cdot}}, Cat^+)$. This reaction proceeds according to the equation,

$$2A^{\bar{\cdot}}, Cat^+ \rightleftarrows A + A^{2-}, 2Cat^+, \qquad K_{disp}$$

and its equilibrium constant is extremely sensitive to changes of solvent and cation.

Disproportionation of radical anions may be investigated by a variety of techniques. Its equilibrium constant can be calcu-lated from the difference between the first and second redox potential of the parent compounds and hence the potentiometric or polarographic techniques are useful for its determination. For the sake of illustration the potentiometric data of Rainis and Szwarc (12) augmented by the polarographic findings of Parker and his student (13) are collected in Table I. They clearly demonstrate the great effect exerted by solvent and cation upon the value of the K_{disp} of anthracene and perylene radical anions.

Even more striking are the results leading to the respective ΔH_{disp} and ΔS_{disp}. By using the device depicted in Figure 6 we measured (14) the temperature dependence of $\Delta\varepsilon$--the difference of the first and second redox potential of the parent compounds. The employed apparatus consists of two containers, one filled with a solution of 1:1 mixture of A and $A^{\overline{\cdot}}$,Cat^+, the other with a solution of 1:1 mixture of $A^{\overline{\cdot}}$,Cat^+ and A^{2-},$2Cat^+$. Two platinum electrodes immersed in the respective solutions are connected to a voltameter of infinite resistence and the electric circuit is completed by having a liquid junction (Cat^+,BPh_4^-) linking the containers. The whole device is placed in a Dewar flask filled with a liquid kept at any desired temperature.

TABLE I

Disproportionation Constant, K_{disp} $2A^{\overline{\cdot}}$,$Cat^+ \rightleftarrows A + A^{2-}$,$2Cat^+$

Cation	Solvent	K_{disp}	Cation	Solvent	K_{disp}
		A = anthracene			
Li^+	DME	1.8×10^{-8}	Li^+	THF	2.9×10^{-7}
Na^+	DME	2.8×10^{-8}	Na^+	THF	1.6×10^{-5}
K^+	DME	1.2×10^{-7}	K^+	THF	6.8×10^{-7}
NBu_4^+	DME	2.3×10^{-9}*	Cs^+	THF	
NBu_4^+	DMF	2.2×10^{-14}*	NBu_4^+	THF	1.6×10^{-13}*
		A = Perylene			
Li^+	DME	3.8×10^{-10}	Li^+	THF	2.0×10^{-9}
Na^+	DME	8.9×10^{-10}	Na^+	THF	1.7×10^{-6}
K^+	DME	8.9×10^{-8}	K^+	THF	2.2×10^{-6}
NBu_4^+	DME*	1.6×10^{-10}	NBu_4^+	THF*	1.6×10^{-10}
NBu_4^+	DMF*	3.9×10^{-11}	Na^+	DMF*	5.9×10^{-11}

*Data from ref. 13; all other data from ref. 12.
    DME - dimethoxyethane; THF - tetrahydrofuran; DMF - dimethyl-formamide.

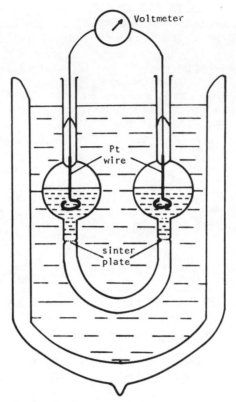

Figure 6.   Schematic drawing of the device measuring the
temperature dependence of the equilibrium constant of disproportionation of radical anions.

The measured potential, $\Delta\varepsilon$, gives the free energy of disproportionation provided that the potential of the liquid junction
is negligible. This seems to be the case. Its value is determined
for a series of temperatures, first by raising the temperature of
the bath from $-70°C$ to $25°C$ and then by lowering it again. Plots
of $\Delta\varepsilon$ versus T and of $\Delta\varepsilon/RT = -\ln K_{disp}$ versus $1/T$ were used to
determine the respective ΔS and ΔH, and, for the sake of illustration, such plots for disproportionation of sodium perylenide in
tetrahydrofuran are shown in Figure 7.

Some of the results obtained in THF for the Li^+, Na^+, K^+ and Cs^+
salts of perylenide are collected in Table II. They confirm the
interpretation proposed by Rainis and Szwarc (12) to account for
the peculiar behavior of the sodium salt. The ΔS values show that

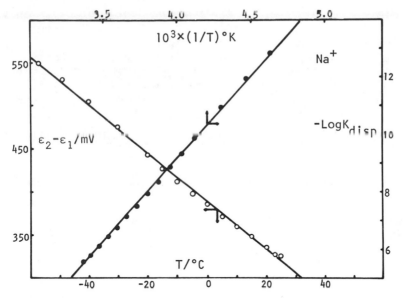

Figure 7.   Plots of $\varepsilon_2 - \varepsilon_1$ versus T and of $-(\varepsilon_2 - \varepsilon_1)/RT = \ln K_{disp}$ versus 1/T for the disproportionation of sodium perylenide in THF.

TABLE II

ΔH and ΔS for the Disproportionation in THF of Perylenide
Salts $(Pe^{\overline{\cdot}}, Cat^+)$

$$2Pe^{\overline{\cdot}}, Cat^+ \rightleftarrows Pe + Pe^{2-}, 2Cat^+; \quad K_{disp.}$$

Cation	temperature range °C	$\Delta H/kcal\ mole^{-1}$	$\Delta S/e.u.$
Li^+	-55 to -40	13.6	0.0
Li^+	0 to +25	16.2	13.1
Na^+	-55 to +25	27.2	65.8
K^+	-55 to +25	14.2	22.2
Cs^+	-55 to +25	11.0	15.1

ΔH from van't Hoff plot: ΔS from a plot $\varepsilon_2 - \varepsilon_1$ versus T.

for the Li^+, K^+ and Cs^+ salts the solvation of cations is only slightly affected by disproportionation, while a large degree of desolvation takes place during the disproportionation of the sodium salt. This indicates that the strongly solvated Li^+ cations retain their solvation shell even on association with dianions. The K^+ and Cs^+ cations are poorly solvated by THF even when free (15) and hence there is little change in the degree of their solvation as they become associated with dianions being previously coupled with radical anions. However, the Na^+ cations are capable of retaining their solvation shell when coupled with perylenide radical ions but on more tight association with dianions a large part of their solvation shell is squeezed out.

Potentiometric or polarographic techniques are useful in disproportionation studies when the respective equilibrium constants K_{disp}, are smaller than 10^{-2}. In fact, they become more reliable for still less favorable disproportionations and K_{disp} as low as 10^{-15} were determined by this technique. Investigation of more favorable disproportionations, i.e., $10^{-2} < K_{disp} < 10^2$ may be accomplished by utilizing spectrophotometric techniques because the absorption spectra of radical anions and dianions are different. For still more favorable disproportionations, ESR technique is useful. It permits a direct determination of the minute concentration of radical anions--the only paramagnetic species present in these systems. This approach permits to determine the disproportionation constant as high as 10^5.

Combination of these techniques was applied in studies of the disproportionation of radical anions of tetracene (15) ($Te^{\bar{}}$,Cat^+). The results summarized in Table III show vividly the enormous effects of solvent and cation. Although the same radical anions are involved in all the systems discussed here, the K_{disp}'s vary from 10^{-14} to 10^3--increasing by 17 orders of magnitude.

The disproportionation of sodium tetracenide in benzene (16) is particularly interesting by revealing the importance of cation's solvation. The K_{disp} in pure benzene could not be determined, most probably its value is greater than 10^6. However, addition of small amounts of THF to the benzene solution favors the formation of $Te^{\bar{}}$,Na^+. The formal disproportionation constant, K_{app}, given by the ratio

$$[Te^{2-},2Na^+][Te]/[Te^{\bar{}},Na^+]^2 = K_{app},$$

is not truly constant, its value depends on the concentration of the added THF. In fact, the plot of log K_{ap} versus log [THF] is linear and has a slope of 2. This implies that the disproportionation in this system results from the reaction

TABLE III

Equilibria and Kinetics of Disproportionation of
Tetracene Radical Anions at 20°C

$$2Te\overline{\cdot},Cat^+ \underset{k_{-1}}{\overset{k_1}{\rightleftharpoons}} Te + Te^{2-},2Cat^+ \qquad K_1$$

Solvent	Cation	K_1	$k_1 \cdot Msec$	$k_{-1} \cdot Msec$
THF	Li^+	5.8×10^{-9}	3.6×10	6.3×10^9
THF	Na^+	1.0×10^{-5}	5.5×10^4	5.5×10^9
THF	K^+	4.6×10^{-6}	3.0×10^4	6.5×10^9
THF	Cs^+	3.2×10^{-6}	2.5×10^4	7.8×10^9
DOX	Li^+	6.6×10^{-2}	6.0×10^6	1.1×10^8
DOX	Na^+	6.5×10^{-2}	2.0×10^7	3.1×10^8
DOX	K^+	1.1×10^{-2}	a	a
DOX	Cs^+	6.5×10^{-3}	a	a
DEE	Li^+	1.64×10	5.9×10^7	3.6×10^6
DEE	Na^+	1.2×10^{-1}	2.4×10^8	2.0×10^9

[a]Flash of visible light ($\lambda < 400$ nm) does not lead to bleaching. Our observation starts 25 μs after flash.

$$2Te\overline{\cdot},Na^+,THF_{n+1} \rightleftharpoons Te + Te^{2-},2(Na^+,THF_n) + 2THF$$

with $\underline{n}$ most probably equal to 1. In other words, the Na^+ cations relatively loosely associated with $Te\overline{\cdot}$ are more extensively solvated by THF than the Na^+ cations more tightly linked to the dianions. The previously mentioned desolvation phenomenon can be now described in a quantitative way.

The association of radical anions and dianions with cations provides the driving force for the disproportionation. In the most simplistic terms, the Coulombic energy of $A^{\bar{\cdot}}, Cat^+$ pairing is given by e^2/r. Providing that the effective distance r between the dianion and cations is the same in the $A^{2-}, 2Cat^+$ aggregate as in the $A^{\bar{\cdot}}, Cat^+$ pair, the disproportionation increases the Coulombic energy of the system by $2e^2/r$ (i.e., $4e^2/r - 2e^2/r$). The gain is even larger because the effective r is smaller in the $A^{2-}, 2Cat^+$ aggregate than in the $A^{\bar{\cdot}}, Cat^+$ pairs.

The effective distance r depends on the size of the anion. This is illustrated by the data collected in Table IV referring to disproportionation of aromatic radical anions in diethyl ether with Li^+ counterions (17). In this system Li^+ is hardly solvated and hence r's are small, thus magnifying the observed effects. An approximately linear relation was found between log K_{disp} and $1/r_{eff}$, confirming the proposed explanation. It should be stressed at this point that disproportionation squeezes two electrons in one orbital, while they were placed in separate orbitals of two $A^{\bar{\cdot}}$. Hence, the electron-electron repulsion increases on disproportionation. In spite of this, K_{disp} is greater for smaller hydrocarbons, a clear indication of the importance of cation-anion interaction.

The importance of ion pairing in the disproportionation is further manifested when the following reactions, taking place in THF, are compared:

$$2T^{\bar{\cdot}}, Na^+ \;\rightleftharpoons\; T + T^{2-}, 2Na^+ \qquad K_1 = 400$$

$$T^{\bar{\cdot}} + T^{\bar{\cdot}}, Na^+ \;\rightleftharpoons\; T + T^{2-}, Na^+ \qquad K_2 = 3.3$$

$$2T^{\bar{\cdot}} \;\rightleftharpoons\; T + T^{2-} \qquad\qquad \text{estimated } K_3 \sim 10^{-8}.$$

In these equations T denotes tetraphenylethylene, and the above data are taken from a paper by Roberts and Szwarc (18). Similar studies involving lithium salts were reported later (19) and showed the same trend in the disproportionation constants.

Thermodynamics of disproportionation of radical anions was discussed in the preceding paragraphs. We turn now to the problems of kinetics. The method used in the kinetic studies is based on the following stratagem:

We, as well as other investigators, demonstrated that a flash of visible light ejects electrons from radical anions or dianions. For example, a flash of visible light leads to electron photoejection from pyrene radical-anions $(\pi^{\bar{\cdot}})$ (20) resulting in the reaction

TABLE IV

Disproportionation Constants K_{disp} of Lithium Salts of Aromatic
Radical Anions in Diethyl Ether at Ambient Temperature

Hydrocarbon	K_{disp}	$2e^2/\varepsilon_{eff} \sim \Delta E$
Biphenylene	1270	0.5574
Naphthalene	1270 ± 120	0.5558
Anthracene	42 ± 7	0.5244
Tetracene	16 ± 1	0.5000
Perylene	0.76 ± 0.11	0.4780
Pyrene	0.27 ± 0.03	0.4948

$$\pi\overline{\cdot} \;\xrightarrow{\;h\nu\;}\; \pi + e^{-}.$$

In the presence of an excess of a hydrocarbon of low electron affinity, e.g. biphenyl B, some of the ejected electrons are cap-tured by B and thus $B\overline{\cdot}$ is formed after a flash by the reaction,

$$B + e^{-} \rightarrow B\overline{\cdot},$$

although the equilibrium

$$\pi\overline{\cdot} + B \;\underset{\leftarrow}{\rightarrow}\; \pi + B\overline{\cdot}$$

lays far to the left. Hence, the above equilibrium is upset by the flash and the system returns to its original state of equilib-rium through the process

$$B\overline{\cdot} + \pi \rightarrow B + \pi\overline{\cdot}.$$

Since $B\overline{\cdot}$ has a strong absorbance at 400 nm while $\pi\overline{\cdot}$ absorbs strongly at 490 nm, the progress of the above electron-transfer reaction can be monitored by the conventional flash-photolysis technique. Thus, the bimolecular rate constant of this electron transfer can be determined from the oscilloscope tracings.

In the presence of excess of Na^{+} ions, the radical anions could be quantitatively converted into ion pairs, and the same approach allows the determination of the bimolecular rate constant

$$B\overline{\cdot},Na^{+} + \pi \rightarrow B + \pi\overline{\cdot},Na^{+}.$$

Electron transfer coupled with Na^{+} transfer was found to be about 10 times slower than the corresponding process involving free ions. Both refer to reactions taking place in THF.

The phenomenon of electron photoejection was also clearly demonstrated in photolysis of sodium perylenide induced by visible light and taking place in the absence of the parent hydrocarbon (21). The difference spectrum recorded at 100 μsec after the flash showed 2 absorption bands at 408 and 437 nm that revealed the formation of perylene, and a bleaching at 580 nm where perylenide absorbs.

Electron photoejection from radical anions may result in capture of the ejected electron by an unphotolyzed radical anion. A net result of such events is a loss of two radical anions, one through the photolysis and the other caused by the electron cap-ture, and formation of one molecule of the parent hydrocarbon and one dianion,* i.e.,

$$A^{\bar{\cdot}},Cat^+ \xrightarrow{h\nu} A + e^-,Cat^+$$

$$A^{\bar{\cdot}},Cat^+ + e^-,Cat^+ \rightarrow A^{2-},2Cat^+$$

Hence, equilibrium in an equilibrated solution of A, $A^{\bar{\cdot}},Cat^+$, and $A^{2-},2Cat^+$ is upset by a flash and the system returns to its equilibrium state in the following dark period by the process

$$A + A^{2-},2Cat^+ \rightleftarrows 2A^{\bar{\cdot}},Cat^+.$$

When the disproportionation is highly favored, i.e., the equilibrium concentration of $A^{\bar{\cdot}},Cat^+$ is much smaller than those of A and $A^{2-},2Cat^+$, the electron photoejection takes place mainly from the dianions and the capture involves mainly the hydrocarbon. Thus,

$$A^{2-},2Cat^+ \xrightarrow{h\nu} A^{\bar{\cdot}},Cat^+ + e^-,Cat^+$$

$$A + e^-,Cat^+ \rightarrow A^{\bar{\cdot}},Cat^+,$$

and two radical anions are formed as one A and one $A^{2-},2Cat^+$ are lost. The disproportionation equilibrium is again upset and the system returns to it during the dark period following the flash by the reaction

$$2A^{\bar{\cdot}},Cat^+ \rightleftarrows A + A^{2-},2Cat^+.$$

In either case the relaxation of the system can be studied by the conventional spectrophotometric techniques, and thus obtained data, coupled with the known equilibrium constant of disproportionation, allow us to determine the rate constants of disproportionation and of its reverse reaction.

The approach described in the preceding paragraphs was applied to a variety of systems (15,19,21) and some of the reported rate data are included in Table III. Two of the investigated systems deserve further comments.

The previously discussed equilibrium studies of sodium tetracenide $(Te^{\bar{\cdot}},Na^+)$ disproprotionation in benzene containing small amounts of tetrahydrofuran (THF) demonstrated that the equilibrium of the disproportionation is described by the equation

---

*Electron ejection from $A^{\bar{\cdot}},Cat^+$ followed by its capture by A leaves the system unaltered.

$$2Te^{-},Na^{+},THF_{n+1} \rightleftharpoons Te + Te^{2-},2(Na^{+},THF_n) + 2THF.$$

This reaction may proceed through two routes, viz (1,b) or (a,2)

$$2Te^{-},Na^{+}THF_{n+1} \underset{k_{-1}}{\overset{k_1}{\rightleftharpoons}} Te + Te^{2-},2(Na^{+},THF_{n+1})$$

(a) $\Big\Updownarrow K_a$ (b) $\Big\updownarrow K_b$

$$2Te^{-},Na^{+}THF_n + 2THF \underset{k_{-2}}{\overset{k_2}{\rightleftharpoons}} Te + Te^{2-},2(Na^{+},THF_n) + 2THF$$

In view of the relatively high concentrations of THF the relaxation times for both solvation-desolvation processes are expected to be much shorter than the relaxation time of (1) or (2). Let us stress also that the solvation equilibrium (a) strongly favors the fully solvated radical ion pairs, whereas the solvation-desolvation equilibrium (b) prefers greatly the partially desolvated dianion aggregates, i.e. K_a is large while K_b is small. This implies that $[Te^{-},Na^{+},THF_{n+1}] \gg [Te^{-},Na^{+},THF_n]$ and $[Te^{2-},2(Na^{+},THF_n)] \gg [Te^{2-},2(Na^{+},THF_{n+1})]$.

For a reaction proceeding through route (1) followed by equilibrium (b) the rate of the forward step should be independent of [THF], being given by $k_1 [Te^{-},Na^{+},THF_{n+1}]^2$, whereas the rate of the backward reaction would be enhanced by increasing the concentration of THF since it is determined by $(k_{-1}/K_b)[Te][Te^{2-}, 2(Na^{+},THF_n)][THF]^2$. On the other hand, for the route arising from the desolvation of radical anion pairs followed by (2), one demands retardation of the forward reaction by increasing THF concentration, its rate being

$$(k_2/K_a)[Te^{-},Na^{+},THF_{n+1}]^2/[THF]^2,$$

whereas the rate of the backward reaction should be independent of [THF], being given by $k_{-2}[Te][Te^{2-},2(Na^{+},THF_n)]$.

Depending on whether k_1 is greater or smaller than $k_2/K_a [THF]^2$, the first or second route will be favored. Of course, if one of the routes is favored for the forward reaction is has to be also favored for the backward reaction since the condition of equilibrium demands $k_1/(k_{-1}/K_b) = (k_2/K_a)/k_{-2}$.

The system discussed here was studied under conditions where concentration of radical anions exceeded by a factor of about 10 the concentration of either Te or its dianion. A flash of light led to increase in radical anion concentration and to about 10% bleaching of the dianions. The return to equilibrium was therefore mainly due to the forward reaction which virtually obeyed the first order law with the pseudo-first order rate constant proportional to $[\mathrm{Te^{\bar{\cdot}}, Na^{+}, THF_{n+1}}]_{\mathrm{equil}}$. and retarded by increase of $[\mathrm{THF}]$. Hence, the reaction proceeds through the route $(a, 2)$, the backward rate-determining step has to be given by $\underline{k_{-2}}\,[\mathrm{Te}][\mathrm{Te^{2-}, 2(Na^{+}, THF)_{n}}]$ whereas the rate determining step of the forward reaction is given by $\underline{k_{2}}[\mathrm{Te^{\bar{\cdot}}, Na^{+}, THF_{n}}]^{2}$. In other words, the less solvated dianion aggregate reacts with Te to yield the not fully solvated radical anion pairs which in turn rapidly incorporate additional THF molecules into their solvation shell. The disproportionation of the fully solvated radical anion pairs cannot compete with the rapid disproportionation of the partially desolvated pairs, in spite of their low concentration.

Disproportionation of barium salt of tetracenide presents another interesting system (22). Under the conditions of our experiments (radical anion concentration $< 10^{-6}\mathrm{M}$) the barium salt of radical anions is virtually fully dissociated

$$\mathrm{Ba^{2+}(Te^{\bar{\cdot}})_{2}} \rightleftarrows \mathrm{Ba^{2+}, Te^{\bar{\cdot}} + Te^{\bar{\cdot}}}.$$

Disproportionation is highly favored and, therefore, the system is composed of Te, $\mathrm{Te^{2-}}$, $\mathrm{Ba^{2+}}$ and of much smaller amounts of radical anions. A flash of light leads to partial bleaching of $\mathrm{Te^{2-}}$, $\mathrm{Ba^{2+}}$ and to electron capture by Te. The equilibrium of disproportionation,

$$\mathrm{Ba^{2+}, Te^{\bar{\cdot}} + Te^{\bar{\cdot}}} \rightleftarrows \mathrm{Te + Ba^{2+}, Te^{2-}},$$

is upset, but the return to its original state may involve two routes, viz.,

$$\mathrm{Ba^{2+}, Te^{\bar{\cdot}} + Te^{\bar{\cdot}}} \rightleftarrows \mathrm{Ba^{2+}, Te^{2-} + Te}$$

$$\mathrm{Te^{\bar{\cdot}}, Ba^{2+}, Te^{\bar{\cdot}}}$$

The direct disproportionation may be visualized as the approach of $\mathrm{Te^{\bar{\cdot}}}$ to $\mathrm{Ba^{2+}, Te^{\bar{\cdot}}}$ from the side of $\mathrm{Te^{\bar{\cdot}}}$, whereas the indirect disproportionation leading first to the undissociated salt, $\mathrm{Te^{\bar{\cdot}}, Ba^{2+}, Te^{\bar{\cdot}}}$ might involve the approach of $\mathrm{Te^{\bar{\cdot}}}$ to $\mathrm{Ba^{2+}}$ side of $\mathrm{Ba^{2+}, Te^{\bar{\cdot}}}$. Kinetic results (22) indicate the participation of both routes in the overall process and estimate of the pertinent rate constant was reported.

I propose to end this review with two examples illustrating the application of this flash-photolytic technique to characterization of labile radical anions and to study of their behavior.

Salts of cis-stilbene radical anions rapidly isomerize into their trans-isomers (23) and hence their optical spectrum cannot be recorded directly like the spectrum of the salts of trans-stilbene radical ions. The isomerization of cis-stilbene is catalyzed by electron transfer (23), but this reaction is imperceptibly slow when perylenide is used as the electron donor. We can mix, therefore, sodium perylenide with an excess of cis-stilbene in THF without inducing any reaction. Flash of light ejects electrons from the perylenide and some of them are subsequently captured by cis-stilbene. The difference spectrum observed in the period 100-300 µsec after a flash shows a transient absorption attributed to cis-stilbene radical anions, and a transient bleaching caused by the disappearance of an equivalent amount of perylenide (24). Since the spectrum and molar extinction of the latter is known, one can calculate the spectrum and molar extinction of the unknown cis-stilbene radical anions from the observed difference spectrum. To ascertain the reliability of the method, analogous experiments were performed with trans-stilbene; the derived spectrum fully agreed with that directly obtained with the aid of a spectrophotometer. It should be stressed that in both systems the absorbance and bleaching disappear after a few hundred µsec; nevertheless the ratios of the absorbance to bleaching, measured at the same time, remained constant and independent of time. The spectra of cis- and trans-stilbenides determined by this method are shown in Figure 8.

Having the molar absorbance of cis- and trans-stilbenides we could determine the equilibrium constant of the electron transfer process:

cis-stilbenide + trans-stilbene $\rightleftarrows$ cis-stilbene + trans-stilbenide.

This was accomplished by flash-photolyzing a mixture of cis- and trans-stilbenes of known composition added to a small amount of sodium perylenide. The ratio r of the absorbance in a region where the stilbenides absorb, e.g., at 500 nm, to bleaching of the perylenide band at 578 nm is given by

$$r \varepsilon_{\text{perylenide at 578}} = X \varepsilon_{\text{cis at 500}} + (1-X) \varepsilon_{\text{trans at 500}} \; .$$

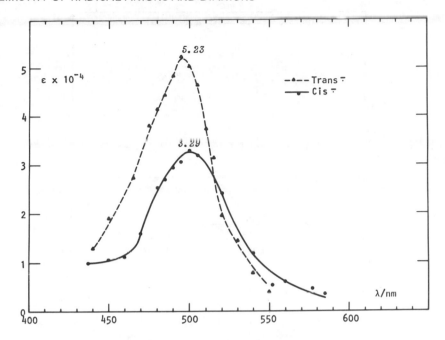

Figure 8. Absorption spectra of sodium salts of <u>cis</u>- and
<u>trans</u>-stilbenides in THF obtained by flash-photolytic technique.

The value of X gives the required equilibrium constant, K_{tr}, equal
to

$$K_{tr} = \{(1-X)/X\}\{[\underline{cis}-stilbene]/[\underline{trans}-stilbene]\}.$$

Thus K_{tr} was found to be 5 for the electron transfer involving
sodium salts in THF (24).

    For the final example we shall consider the flash-photolysis
approach to studies of radical anions of 1,1-diphenylethylene (D).
These again are highly labile radical anions because they dimerize
rapidly into stable dimeric dianions, $Ph_2\bar{C}.CH_2CH_2.\bar{C}Ph_2$ ($^-DD^-$).
Flash-photolysis of $Na^+,^-DD^-,Na^+$ in the presence of an excess of
D leads to dissociation of the dimer into $2D^{\bar{\cdot}},Na^+$, presumably
through electron ejection. The transient bleaching of the absorption of $Na^+,^-DD^-,Na^+$ compared with the appearance of the transient
absorbance allows us to construct the absorption spectrum of the
labile $D^{\bar{\cdot}},Na^+$ shown in Figure 9 (25).   The system returns to its
original state during the dark period following the flash.   All
the transients simultaneously disappear, their absorbances decreasing according to the bimolecular law, i.e., $1/\Delta(od)$'s are linear

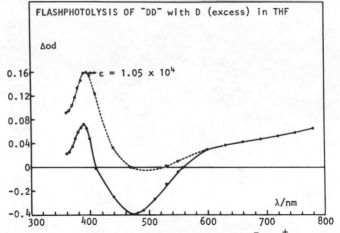

Figure 9.   Absorption spectrum of $D^{\bar{\cdot}},Na^+$ in THF.

with time. Since the molar absorbance $Na^+,^-DD^-,Na^+$ is determined, the bimolecular rate constant of $D^{\bar{\cdot}},Na^+$ dimerization could also be determined (25) (5×10^8 M^{-1} sec^{-1}).

Flash-photolysis of $Na^+,^-DD^-,Na^+$ performed in the presence of a mixture of D and triphenylene (Trph) yields the transient absorbance of triphenylide (Trph$^{\bar{\cdot}},Na^+$). The return of the system to its original state is now much slower but still is governed by the second order law. Obviously, the equilibrium

$$D^{\bar{\cdot}},Na^+ + Trph \rightleftarrows D + Trph^{\bar{\cdot}},Na^+, \qquad K_{D,Trph}$$

decreases the concentration of $D^{\bar{\cdot}},Na^+$ and retards the dimerization. Thus, $2\underline{d\Delta}[Na^+,^-DD^-,Na^+]/dt = d\{[D^{\bar{\cdot}},Na^+] + [Trph^{\bar{\cdot}},Na^+]\}/dt = \underline{k_d}$ $[D^{\bar{\cdot}},Na^+]^2$. It follows that $1/\underline{\Delta}$ od (470), i.e., the reciprocal of the absorbance of bleached $Na^+,^-DD^-,Na^+$ is given by

$$1/\Delta od(470) = \underline{2k_d t}.\varepsilon/(1 + K_{D,Trph}[Trph]/[D])^2,$$

i.e., the square root of the reciprocals of the slopes of lines giving $1/\Delta od(470)$ vs. $\underline{t}$, should be linear with the ratio [Trph]/ [D]. This was indeed the case and from the slope of such a line $K_{D,Trph}$ was calculated. Since the redox potential of Trph is known, our approach led to determination of the redox potential of D resulting in formation of a labile $D^{\bar{\cdot}}$.

In conclusion, I wish to thank the National Science Foundation for their continuous support of all these studies.

REFERENCES

(1) "Ions and Ion Pairs in Organic Reactions," M. Szwarc, Ed.,
Wiley, 1972.

(2) N. M. Atherton and S. I. Weissman, J. Am. Chem. Soc., 83,
1330 (1961).

(3) K. Hoffelmann, J. Jagur-Grodzinski and M. Szwarc, J. Am.
Chem. Soc., 91, 4645 (1969).

(4a) N. Hirota, J. Am. Chem. Soc., 90, 3603 (1968).

(4b) N. Hirota, R. Caraway and W. Schook, Ibid., 3611 (1968).

(5) T. E. Gough and P. R. Hindle; Can. J. Chem., 47, 1698, 3393
(1969); Trans. Faraday Soc., 66, 2420 (1970).

(6) T. E. Hogen-Esch and J. Smid, J. Am. Chem. Soc., 88, 307,
318 (1966).

(7a) K. H. Wong, G. Konizer and J. Smid, J. Am. Chem. Soc., 92,
666 (1970).

(7b) U. Takaki, T. E. Hogen-Esch and J. Smid, ibid., 93, 6760
(1971).

(8a) T. E. Hogen-Esch and J. Smid, J. Am. Chem. Soc., 91, 4580
(1969).

(8b) J. Smid, Ang. Chem., 84, 127 (1972).

(9) G. Levin, B. Lundgren, M. Mohammad and M. Szwarc, J. Am.
Chem. Soc., 98, 1461 (1976).

(10) B. De Groof, G. Levin and M. Szwarc; J. Am. Chem. Soc., 99
474 (1977).

(11a) R. V. Slates and M. Szwarc, J. Am. Chem. Soc., 89, 6043
(1967).

(11b) L. Lee, R. Adams, J. Jagur-Grodzinski and M. Szwarc, J. Am.
Chem. Soc., 93, 4149 (1971).

(12) A. Rainis and M. Szwarc, J. Am. Chem. Soc. 96, 3008 (1974).

(13) R. S. Jensen and V. D. Parker, J. Am. Chem. Soc., 97, 5619
(1975).

(14) H. C. Wang, G. Levin and M. Szwarc, J. Am. Chem. Soc., in press (1977).

(15) G. Levin and M. Szwarc, J. Am. Chem. Soc., 98, 4211 (1976).

(16) J. Pola, G. Levin and M. Szwarc, J. Phys. Chem., 80, 1690 (1976).

(17) G. Levin, B. E. Holloway and M. Szwarc, J. Am. Chem. Soc. 98, 5706 (1976).

(18) R. C. Roberts and M. Szwarc, J. Am. Chem. Soc. 87, 5542 (1965).

(19) B. Lundgren, G. Levin, S. Claesson and M. Szwarc, J. Am. Chem. Soc. 97, 262 (1975).

(20) G. Rämme, M. Fisher, S. Claesson and M. Szwarc, Proc. Roy. Soc. A. 327, 467 (1972).

(21) G. Levin, S. Claesson and M. Szwarc, J. Am. Chem. Soc. 94, 8672 (1972).

(22) B. De Groof, G. Levin and M. Szwarc, J. Am. Chem. Soc. 99, 474 (1977).

(23) T. A. Ward, G. Levin and M. Szwarc, J. Am. Chem. Soc. 97, 258 (1975).

(24) H. C. Wang, G. Levin and M. Szwarc, J. Am. Chem. Soc., 99, 2642 (1977).

(25) H. C. Wang, E. D. Lillie, S. Slomkowski, G. Levin and M. Szwarc, J. Am. Chem. Soc., in press.

MECHANISMS OF NUCLEOPHILIC ADDITIONS TO CATION RADICALS

OF EE SYSTEMS

Henry N. Blount and John F. Evans

Brown Chemical Laboratory

The University of Delaware

Newark, Delaware 19711

Reactions of cation radicals with various electron-rich species have been the subject of many recent literature reports (1-35) and reviews (36-41). A major focal point of these studies has been the investigation of reactions of nucleophiles and reducing agents with cation radicals derived from substrates which are capable of undergoing multiple, single-electron oxidation processes within the potential window imposed by the reaction environment:

$$A \rightleftharpoons A^{+\cdot} + e^- \qquad E_1 \tag{1}$$

$$A^{+\cdot} \rightleftharpoons A^{++} + e^- \qquad E_2 \tag{2}$$

where

$$E_1 < E_2 \tag{3}$$

Interest in these "EE" systems stems from the thermodynamically dictated disproportionation process

$$2A^{+\cdot} \underset{k_b}{\overset{k_f}{\rightleftharpoons}} A^{++} + A \tag{4}$$

which predicts the formation of the dicationic form of the substrate in solutions containing the cation radical. Although the equilibrium constants for these disproportionation processes

$$K_{dis} = \frac{k_f}{k_b} = \exp\left[-\frac{nF}{RT}(E_2 - E_1)\right] \tag{5}$$

are generally small (11,33,42,43), the attendant kinetics can be quite fast (44-46) thereby providing an avenue for cation radical reaction through the more electrophilic dication. The extent to which this process contributes to the observed kinetics of cation radical/nucleophile reactions must always be considered in the study of cation radicals derived from EE substrates.

Recent work in these laboratories has been directed toward gaining a mechanistic understanding of the reactions of cation radicals of EE systems with nucleophilic species and toward a systematic evaluation of the impact of both the nature of the nucleophile and the structure of the substrate on these mechanisms. In their reactions with a common carbon-centered cation radical, that derived from 9,10-diphenylanthracene (DPA), eleven nucleophiles of varying classical nucleophilicity (47) and redox potential (48-51) were examined. These nucleophiles were categorized according to their structural features as (I) Anionic nucleophiles, (II) Aprotic neutral nucleophiles and (III) Protic neutral nucleophiles as shown in Table I. There were three objectives of this portion of the work: (1) To characterize the dynamics and assign mechanisms to the reactions of these nucleophiles with the cation radical of DPA (DPA$^{+\cdot}$), (2) To delineate where possible those properties of these nucleophiles which determine whether they function as reducing agents or undergo addition reactions with this given electrophile, and (3) To establish a scale of relative reactivities of all nucleophiles examined in this study without regard to reaction type (addition or electron transfer).

With regard to the effect of substrate structure on the reactions of nucleophiles with cation radicals, the influence exerted by the presence of heteroatoms on the mechanism operative in nucleophilic additions to sulfur-containing, fused-ring cation radicals derived from EE substrates was probed. The reactions of thianthrene (TH) cation radical (TH$^{+\cdot}$) with water and pyridine were selected as exemplary cases from which comparisons with the analogous DPA$^{+\cdot}$ reactions were drawn. Since water was known to add to TH$^{+\cdot}$ at the sulfur heteroatom (1,2), whereas pyridine had been shown to yield a ring substitution product (4,6), reactions of TH$^{+\cdot}$ with these nucleophiles allowed direct comparison of the two processes. Furthermore, the reaction of 10-phenylphenothiazine (PH) cation radical (PH$^{+\cdot}$) with pyridine provided insight into the effect of the presence of a nitrogen heteroatom on the dynamics and mechanism of pyridination.

TABLE 1

Redox Potentials and Nucleophilicities of Nucleophiles
Examined in Reaction with $DPA^{+\cdot}$

Nucleophile[a]	$E_{1/2}$, V[b]	(Ref.)	Nucleophilicity[c], N/CH_3OH n_{CH_3I,CH_3OH}	(Ref.)
I^-	0.50	51	7.42	64
CN^-	< 0.60	20	6.70	64
SCN^-	0.69	51	6.70	64
Br^-	0.85	51	5.79	64
Cl^-	1.11	51	4.37	64
Et_3N	1.7	50	6.66	64
Py	2.3	50	5.23	64
CNPy	> 2.5	d	3.76	67
H_2S	0.14	63	3.34	65
H_2O	1.6	50	1.35	66
Pip	1.7	50	7.30	64

[a] Et_3N = triethylamine, Py = pyridine, CNPy = 4-cyanopyridine, Pip = piperidine.

[b] Voltammetric half-wave potentials (vs SCE) for oxidation at a platinum electrode in CH_3CN.

[c] As defined by Swain and Scott (69) for the displacement of I^- from CH_3I in CH_3OH (64).

[d] This work.

The results presented here point to the importance of the aforementioned considerations regarding both nucleophile and substrate and suggest the broad applicability of a half-regeneration mechanism (9-12,17,19,24,27,30,31,34,52) as a tenable description of the reaction demeanor of these systems.

1. EXPERIMENTAL

Materials

The sources of and purification procedures for acetonitrile and butyronitrile have been reported (11) as have those for pyridine (Py) and 10-phenylphenothiazine (11), DPA (53), TH (10), tetraethylammonium perchlorate (TEAP) and hydrogen sulfide (54), tetra-n-butylammonium perchlorate (TBAP) and tetra-n-butylammonium chloride (TBAC) (34) and triethylamine (33). Preparative procedures have been reported for N-(2-thianthrenyl)-pyridinium perchlorate (TH(Py)$^+$ClO$_4^-$) and thianthrene-5-oxide (TH(O)) (10) and N-[2-(10-phenylphenothiaziny1)]- pyridinium perchlorate (P(Py)$^+$ClO$_4^-$) and 10-phenylphenothiazine-5-oxide (PH(O)) (11).

4-Cyanopyridine (CNPy) (Aldrich) was twice recrystallized from benzene, mp 79-79.5°. Piperidine (Pip) (MCB Manufacturing Chemists, reagent grade) was fractionally distilled from KOH pellets at atmospheric pressure, bp 105-105.5°. Tetra-n-butylammonium iodide (TBAI) (Eastman) was doubly recrystallized from water, pulverized and dried in vacuo at 60° for 6 h., mp 145.5-147°. Tetra-n-butylammonium bromide (TBABr) (Eastman) was twice crystallized from absolute ethanol by the addition of anhydrous ether, vacuum dried at 80° for 6 h., mp 118-119.5°.

Tetraethylammonium cyanide (TEACN) was prepared by the metathesis of sodium cyanide and tetraethylammonium chloride in absolute methanol (55,56). Following recrystallization from acetonitrile, stock solutions of TEACN in acetonitrile were prepared and analyzed by potentiometric titration with aqueous silver nitrate. Tetra-n-butylammonium thiocyanate (TBASCN) was isolated from the metathesis of TBABr and potassium thiocyanate in acetone (57) and vacuum dried at 70° for 24 h., mp 123.5-125°.

Apparatus and Techniques

Kinetic studies were conducted using single potential step spectroelectrochemical (SPS/SE) (58) and stopped-flow (SF) (34,35) techniques. SPS/SE measurements were made using platinum optically transparent electrodes (OTE's) (59) fitted to cells of a previously reported design (60). In all cases conventional single-beam transmission configurations were employed (58). Potential step perturbations were applied to these OTE's via a three-electrode potentiostat equipped with circuitry for the compensation of solution resistance (61).

The spectroelectrochemical technique is not directly applicable to those cation radical/nucleophile systems in which the formal electrode potential of the nucleophile is less anodic than that of the cation radical precursor (62). Such systems preclude in situ electrochemical generation of the cation radical and require either the synthesis of a stable, isolable cation radical salt (10,11) or the external electrogeneration of cation radical (10,34) for kinetic studies by SF or other conventional mixing techniques (11).

Both spectroelectrochemical and stopped-flow kinetic spectrometers were interfaced to a dedicated computer system for the acquisition and reduction of data (10,34,35).

For the reactions of DPA$^{+\cdot}$ with various nucleophiles, the stoichiometries of the reactions were evaluated spectrophotometrically (19) using the stopped-flow apparatus (34). The stoichiometries of the reactions of TH$^{+\cdot}$ with H$_2$O and Py and of PH$^{+\cdot}$ with Py are detailed elsewhere (10,11).

All electrode potentials are reported relative to the aqueous saturated calomel electrode (SCE).

2. RESULTS

In reaction with the carbon-centered cation radical DPA$^{+\cdot}$, the eleven nucleophiles listed in Table I afforded two product types: electron transfer and addition. Attendant to electron transfer reactions is 100% regeneration of the radical ion precursor, while addition reactions give rise to 50% regeneration of this entity. The stoichiometries of these reactions detailed in Table II show that both product types are formed by reactions of both protic and aprotic nucleophiles with DPA$^{+\cdot}$.

In all cases in which electron transfer products are formed, the reaction dynamics are described by a rate law which is first order in cation radical concentration, first order in nucleophile concentration and independent of the concentration of precursor. These same concentration dependencies are valid for addition reactions as well, with the exception of the reactions of DPA$^{+\cdot}$ with Cl$^-$ and Et$_3$N wherein the observed reaction rate is second order with respect to the radical ion. For the general rate law given by equation 6, Table III summarizes the kinetic parameters ascertained for these reactions.

$$\text{Rate} = -\frac{d[DPA^{+\cdot}]}{dt} = k_{obs}[DPA^{+\cdot}]^{\alpha}[\text{Nucleophile}]^{\beta}[DPA]^{\gamma} \qquad (6)$$

The effects of alterations of substrate structure on the reactions of EE cation radicals with selected common nucleophiles were examined by a comparative study of the reactions of the cation radicals of DPA, TH and PH with H_2O and Py. The reactions of the sulfur-centered cation radicals with Py were found to be quite medium dependent as reflected in the stoichiometries of these systems detailed in Table IV.

The reaction of H_2O with $DPA^{+\cdot}$ can be described by a simple rate law which shows the rate of disappearance of cation radical to be first order with respect to the concentrations of both cation radical and nucleophile and independent of the concentration of radical ion precursor. In contrast, the reaction of H_2O with the sulfur-centered cation radical $TH^{+\cdot}$ is far more complex as evidenced by the rate of disappearance of $TH^{+\cdot}$ being second order with respect to the concentration of cation radical, third order with respect to concentration of nucleophile, independent of concentration of precursor, and inversely dependent upon proton concentration. The characteristic kinetic parameters of these two systems are shown in Table V.

On the basis of the observed rate laws for the reactions of Py with $DPA^{+\cdot}$, $TH^{+\cdot}$ and $PH^{+\cdot}$, widely differing reaction pathways are suggested. Together with the observed kinetic parameters for these reactions, the dependencies of the rates of radical ion consumption on the concentrations of radical ion, nucleophile and precursor are summarized in Table VI.

3.   DISCUSSION

Early investigations of the hydrolyses of cation radicals derived from various EE substrates revealed similar product distributions but different rate laws for cation radical consumption. For the hydrolysis of the cation radical of phenoxathiin ($PHOT^{+\cdot}$) in acetonitrile, Cauquis, et al., (12) reported the formation of phenoxathiin-5-oxide [PHOT(O)] and the reformation of PHOT, each product accounting for 50% of the cation radical consumed. Similarly, Sioda (19) reported that the hydrolysis of $DPA^{+\cdot}$ yielded $DPA(OH)_2$ and DPA, 50% of the consumed cation radical being accounted for by each product. The rate of the reaction was found to be first order with respect to the concentrations of both cation radical and H_2O and independent of DPA concentration. Shine and co-workers (1,2) reported that, like $PHOT^{+\cdot}$, the hydrolysis of $TH^{+\cdot}$

TABLE II

Stoichiometries of Reactions of $DPA^{+\cdot}$ with
Various Nucleophiles[a]

Nucleophile	Class[b]	% DPA Regenerated	Reaction Type[c]	Observed Products[d]	References
H_2S	III	97($\pm$3)[e]	ET	$S_1 - S_8$	35
Br^-	I	95($\pm$4)	ET	Br_2	70
I^-	I	98($\pm$3)	ET	I_3^-	70
CN^-	I	97($\pm$5)	ET		70
SCN^-	I	98($+$3)	ET	$(SCN)_x$	70
H_2O	III	51($\pm$1)	AD	$DPA(OH)_2$	19,25,70
Cl^-	I	53($\pm$3)	AD	$DPA(Cl)_2$	19,34
Py	II	50($\pm$2)	AD	$DPA(Py)_2^{++}$	33
CNPy	II	51($\pm$4)	AD		70
Et_3N	II	50($\pm$3)	AD	$DPA(Et_3N)_2^{++}$	33
Pip	III	54($\pm$5)	AD		70

[a] Cation radical electrogenerated from DPA at Pt anode in CH_3CN.

[b] See text.

[c] ET = electron transfer, AD = addition.

[d] Other than DPA; $DPA(OH)_2$ = 9,10-dihydroxy-9,10-diphenyl-9,10-dihydroanthracene; $DPA(Cl)_2$ = 9,10-dichloro-9,10-diphenyl-9,10-dihydroanthracene; $DPA(Py)_2^{++}$ = 9,10-diphenyl-9,10-dipyridinium-9,10-dihydroanthracene; $DPA(Et_3N)_2^{++}$ = 9,10-diphenyl-9,10-di(triethylammonium)-9,10-dihydroanthracene.

[e] Parentheses contain one standard deviation.

TABLE III

Kinetic Parameters for the Reactions of $DPA^{+\cdot}$
With Various Nucleophiles in Acetonitrile
at $25°C^a$

Nucleophile	α	β	γ	k_{obs}, $\underline{M}^{-1}sec^{-1}$	Technique[b]	References
H_2O	1	1	0	$1.29(\pm0.06)^c \times 10^{-1}$	SE,SF	24,34,70
H_2S	1	1	0	$1.32(\pm0.24) \times 10^1$	SE,SF	35
CNPy	1	1	0	$1.41(\pm0.09) \times 10^2$	SE	70
Py	1	1	0	$3.72(\pm0.76) \times 10^4$	SE	31,70
Br^-	1	1	0	$6.91(\pm0.85) \times 10^5$	SF	70
SCN^-	1	1	0	$3.35(\pm0.23) \times 10^6$	SF	70
CN^-	1	1	0	$6.3(\pm1.2) \times 10^6$	SF	70
I^-	1	1	0	$1.49(\pm0.55) \times 10^7$	SF	70
Pip	1	1	0	$2.54(\pm0.36) \times 10^7$	SF	70
Cl^-	2	1	0	$9.26(\pm0.62) \times 10^{9d}$	SF	34,70
Et_3N	2	1	0	$8.0(\pm2.8) \times 10^{13d}$	SF	70

[a] Kinetics monitored for at least two half-lives; TBAP present at 0.10 $\underline{M}$ in all cases.

[b] SE = single potential step spectroelectrochemistry, SF = stopped-flow kinetic spectrophotometry.

[c] Parentheses contain one standard deviation.

[d] Termolecular rate constant, $\underline{M}^{-2} sec^{-1}$.

TABLE IV

Stoichiometries of the Reactions of $DPA^{+\cdot}$, $TH^{+\cdot}$, and $PH^{+\cdot}$ with H_2O and Py

Reaction System	Solvent	% Precursor Regenerated	Other Products[a]	Cation Radical Source[b]	References
$DPA^{+\cdot}/H_2O$	CH_3CN	51	$DPA(OH)_2$	EG	19,25,70
$DPA^{+\cdot}/Py$	CH_3CN	50	$DPA(Py)_2^{++}$	EG	33
$TH^{+\cdot}/H_2O$	CH_3CN	~50	$TH(O)$ (~50%)	CRS	1,2,70
$TH^{+\cdot}/Py$	CH_3NO_2	49	$TH(O)$ (38%) $TH(Py)^+$ (9%)	CRS	4,6
$TH^{+\cdot}/Py$	Py	52	$TH(O)$ (28%) $TH(Py)^+$ (19%)	CRS	4,6
$TH^{+\cdot}/Py$	$CH_3CN/TFAn^c$	48	$TH(O)$ (50%)	CRS	10
$PH^{+\cdot}/H_2O$	CH_3CN	~50	$PH(O)$ (~50%)	CRS	11
$PH^{+\cdot}/Py$	CH_3CN	51	$PH(O)$ (13%) $P(Py)^+$ (35%)	CRS	11,70
$PH^{+\cdot}/Py$	Py	49	$P(Py)^+$ (50%)	CRS	11

[a]Parenthetical values are yields based on quantity of cation radical consumed.
[b]EG = electrogenerated; CRS = cation radical salt (perchlorate).
[c]Acetonitrile containing 4% (v/v) trifluoroacetic anhydride (TFAn).

TABLE V

Kinetic Parameters for the Hydrolyses of DPA$^{+\cdot}$ and TH$^{+\cdot}$ in Acetonitrile at 25°C^a

Cation Radical	Reaction Order With Respect to^b				k_{obs}	Technique	References
	$[A^{+\cdot}]$	$[H_2O]$	$[A]$	$[H_3O^+]$			
DPA$^{+\cdot}$	1	1	0		$1.29(\pm0.06) \times 10^{-1} \underline{M}^{-1} s^{-1}$	SE,SF	24,34,70
TH$^{+\cdot}$	2	3	0	-1	$4.30(\pm0.96) \times 10^3 \underline{M}^{-3} s^{-1}$	SF	10

^{a}Radical ion sources: DPA$^{+\cdot}$ (EG); TH$^{+\cdot}$ (EG, CRS).

^{b}A = radical ion precursor; A$^{+\cdot}$ = radical ion.

TABLE VI

Kinetic Parameters for the Reactions of Pyridine with
$DPA^{+\cdot}$, $TH^{+\cdot}$, and $PH^{+\cdot}$ at 25°C[a]

Cation Radical	Reaction Order With Respect to			k_{obs}	Solvent	Technique	References
	$[A^{+\cdot}]$	$[Py]$	$[A]$				
$DPA^{+\cdot}$	1	1	0	$3.72(\pm0.76) \times 10^{4} \underline{M}^{-1} s^{-1}$	CH_3CN	SE	31,70
$TH^{+\cdot}$	2	1	0	$8.31(\pm0.62) \times 10^{8} \underline{M}^{-2} s^{-1}$	$CH_3CN/TFAn$[b]	SF	10
$PH^{+\cdot}$	2	1	-1	$9.48(\pm0.32) \underline{M}^{-1} s^{-1}$	Py[c]	SF	11

(a) Radical ion sources: $DPA^{+\cdot}$ (EG); $TH^{+\cdot}$ (EG, CRS); $PH^{+\cdot}$ (CRS).

(b) Acetonitrile containing 4% (v/v) TFAn.

(c) Carried out in neat Py and CH_3CN/Py mixtures, see ref. 11.

gave rise to equimolar quantities of TH(0) and TH. The rate of consumption of $TH^{+\cdot}$ was second order in cation radical concentration, first order in H_2O concentration and inversely dependent upon the concentration of TH.

For the hydrolyses of $PHOT^{+\cdot}$ and $DPA^{+\cdot}$, Cauquis, et al., (12) and Sioda (19) both offered the "half-regeneration mechanism" (HRM) as being a tenable pathway for the hydrolyses of these radical ions. For the general cation radical $A^{+\cdot}$, this mechanism may be represented as

$$A^{+\cdot} \;+\; H_2O \;\xrightarrow{\;k\;}\; A(OH)^{\cdot} \;+\; H^{+} \tag{7}$$

$$A(OH)^{\cdot} \;+\; A^{+\cdot} \longrightarrow A(OH)^{+} \;+\; A \tag{8}$$

$$A(OH)^{+} \longrightarrow A(0) \;+\; H^{+} \tag{9a}$$

$$A(OH)^{+} \;+\; H_2O \longrightarrow A(OH)_2 \;+\; H^{+} \tag{9b}$$

with equations (9a) and (9b) accounting for the formation of PHOT(0) and $DPA(OH)_2$, respectively. For the $DPA^{+\cdot}$ system, the initial encounter between cation radical and H_2O is rate determining and gives rise to the aforementioned rate law (24)

$$-\frac{d[A^{+\cdot}]}{dt} = k_{obs}[A^{+\cdot}][H_2O] \tag{10}$$

Dication formation via disproportionation of the cation radical was invoked (1,2) to account for the hydrolysis of $TH^{+\cdot}$. In general form, this mechanism is given by

$$2A^{+\cdot} \;\underset{}{\overset{K_{dis}}{\rightleftharpoons}}\; A \;+\; A^{++} \tag{4}$$

$$A^{++} \;+\; H_2O \;\xrightarrow{\;k\;}\; A(0) \;+\; 2H^{+} \tag{11}$$

wherein a rapid disproportionation equilibrium precedes the rate determining encounter between dication and H_2O, affording the rate law:

$$-\frac{d[A^{+\cdot}]}{dt} = k_{obs}\frac{[A^{+\cdot}]^2[H_2O]}{[A]} \tag{12}$$

Common to both mechanisms is the role of the cation radical as an oxidant. The distinction is that in disproportionation, the

species which is oxidized is another cation radical while in the
half-regeneration scheme, the reductant is the product of the
reaction between the cation radical and the nucleophile (i.e., the
cation radical-nucleophile adduct). In the former case, a rate
law which is second order in cation radical concentration, first
order in nucleophile concentration and inversely dependent upon
precursor concentration is to be expected so long as the reaction
given in equation (11) is rate determining. As the rate of the
reaction between dication and nucleophile increases and kinetic
control is shifted to the forward step of the disproportionation
process [equation (4)], the appropriate rate law becomes independent of both nucleophile and precursor concentrations, but
retains its second order dependence on cation radical concentration (28).

        In the case of half-regeneration [equations (7)-(9)] reaction
kinetics which are first order in concentrations of $A^{+\cdot}$ and H_2O
are observed if and only if the rate of oxidation of the adducted
cation radical [equation (8)] is fast compared to the rate of
adduct formation [equation (7)]. For systems in which the electron
transfer process [equation (8)] is sufficiently slow relative to the
rate of adduct formation [equation (7)] such that the reversible
nature of this latter process is evident (i.e., electron transfer
is rate determining), the appropriate rate law takes the form

$$ - \frac{d[A^{+\cdot}]}{dt} = k_{obs}[A^{+\cdot}]^2[H_2O] \qquad\qquad (13)$$

Furthermore, if the reaction given by equation (8) has equilibrium
character, the disappearance of cation radical is inhibited by
precursor and the observed rate law becomes essentially indis-
tinguishable from that describing the disproportionation pathway.

        Under conditions of in situ electrogeneration of the cation
radical, oxidation of the cation radical-nucleophile adduct may
proceed heterogeneously (22) rather than by the reaction given
in equation (8). Exclusive heterogeneous oxidation of the
adducted cation radical is commonly referred to as an ECE process.
The extent to which this electron transfer step proceeds in concert
with that given in equation (8) is determined by the rate of
heterogeneous electron transfer (71).

        From a thermodynamic point of view, the relative propensity
of a system to react by the HRM vis-a-vis the disproportionation
mechanism is governed by the reversible oxidation potential of the
adducted cation radical

$$A(N)^{+\cdot} \rightleftharpoons A(N)^{++} + e^- \qquad E_3 \qquad\qquad (14)$$

where N is a general nucleophile. In cases where oxidation of the adducted cation radical is more difficult than oxidation of the non-adducted cation radical [equation (2)] (i.e., $E_3 > E_2$), disproportionation is thermodynamically favored. If, as normally expected, oxidation of the adduct is less difficult than oxidation of the cation radical (i.e., $E_3 < E_2$), then the HRM is energetically more favorable. Without regard to these equilibrium thermodynamic considerations, however, heterogeneous and homogeneous kinetic constraints may well dictate the dominant reaction pathway.

Previous work in these laboratories has reaffirmed the applicability of the HRM to the reaction of $DPA^{+\cdot}$ with H_2O (24). This mechanism has also been shown to be operative in the case of the pyridination of $DPA^{+\cdot}$ (31) and recent work has extended the scope of this pathway to include the reactions of Pip and CNPy with this radical ion (70). These systems all reflect 50% regeneration of DPA and adhere to rate laws which are of the form given by equation (10). Evidence for kinetic control by the homogeneous electron transfer step within the HRM [equation (8)] has been provided in the case of the reactions of $DPA^{+\cdot}$ with Cl^- (34) and Et_3N (70). This system affords addition products and reacts according to a rate expression of the form shown in equation (13).

A number of nucleophiles have been examined which, in reaction with $DPA^{+\cdot}$, undergo electron transfer giving rise to 100% regeneration of DPA. In reactions of this cation radical with H_2S (35), SCN^-, I^-, Br^- and CN^- (70), the rate laws for these reactions have been found to take the form of equation (10) indicating rate determining initial encounter between cation radical and nucleophile.

It is noteworthy that a number of nucleophiles act as reducing agents toward $DPA^{+\cdot}$ while others form addition products. Such observations prompt questions regarding the predictability of the product distributions arising from the reactions of the carbon-centered cation radical $DPA^{+\cdot}$ with various nucleophiles. As seen from Tables I and II, simple correlation with redox potential provides guidance for the prediction of reaction type, but does not serve to account for the relative rates of reaction observed.

In this regard, the relative reactivities of those nucleophiles given in Table III (excepting Cl^- and Et_3N) are seen to parallel the corresponding classical nucleophilicities reported by Pearson, et al. (64), for the S_N2 displacement of I^- from CH_3I in CH_3OH given in Table I. We have chosen to define the nucleophilicity parameter for these nucleophiles (N) in reaction with $DPA^{+\cdot}$ in this solvent as

$$n^{N/H_2O}_{DPA^{+\cdot},CH_3CN} = \log\left[\frac{k_N}{k_{H_2O}}\right] \tag{15}$$

where k_N is the bimolecular rate constant for the rate determining
reaction between N and $DPA^{+\cdot}$, and k_{H_2O}, the reference parameter,
is the bimolecular rate constant for the rate determining reaction
between H_2O and $DPA^{+\cdot}$. Stoichiometry dictates that the bimolecular
rate constants for the reaction of $DPA^{+\cdot}$ with H_2O, H_2S, CNPy, Py,
and Pip are numerically one-half of the value of the observed rate
constants shown in Table III. The elementary rate constants and
the corresponding nucleophilicity parameters are shown in Table
VII. Comparison of these $n^{N/H_2O}_{DPA^{+\cdot},CH_3CN}$ values with the $n^{N/CH_3OH}_{CH_3I,CH_3OH}$
parameters given in Table I affords the correlation shown in
Figure 1. If similar transition states were involved in both
electron transfer and addition reactions, then the goodness of fit
evident in Figure 1 can be expected. It follows, then, that the
observed reaction type is a manifestation of the relative stabilities of the products of these initial rate determining encounters
between cation radical and nucleophile (70).

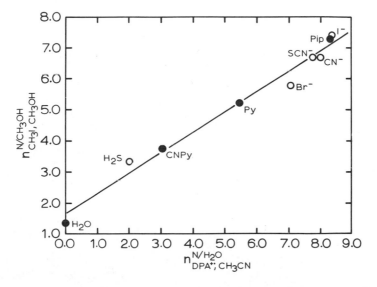

Figure 1. Correlation of nucleophilicity parameters for
reactions with methyl iodide in methanol and with $DPA^{+\cdot}$ in
acetonitrile. Addition cases: solid circles, electron transfer
cases: open circles. Coefficient of correlation = 0.991.

TABLE VII

Elementary Bimolecular Rate Constants[a] and Nucleophilicity
Parameters[b] For the Reactions of Nucleophiles with $DPA^{+\cdot}$

Nucleophile, N	k, $\underline{M}^{-1}s^{-1}$	$n^{N/H_2O}_{DPA^{+\cdot},CH_3CN}$
H_2O	$6.45(\pm0.28)^{c} \times 10^{-2}$	0.00
H_2S	$6.6(\pm1.2)$	2.01
CNPy	$7.03(\pm0.47) \times 10^{1}$	3.04
Py	$1.86(\pm0.38) \times 10^{4}$	5.46
Br^{-}	$6.91(\pm0.85) \times 10^{5}$	7.03
SCN^{-}	$3.35(\pm0.23) \times 10^{6}$	7.72
CN^{-}	$6.3(\pm1.2) \times 10^{6}$	7.99
Pip	$1.26(\pm0.18) \times 10^{7}$	8.29
I^{-}	$1.49(\pm0.55) \times 10^{7}$	8.36

[a] For the rate determining encounter between N and $DPA^{+\cdot}$.

[b] As defined by equation (15).

[c] Parentheses contain one standard deviation.

The oxidative electrochemistry of TH is extremely similar to
that of DPA. The stationary electrode voltammetric peak potentials
corresponding to the formation of the cation radical and dication
[equations (1) and (2)] in acetonitrile are 1.25 V and 1.60 V for
DPA and 1.27 V and 1.68 V for TH. Although the reactivities of the
dications of these systems cause the observed peak potentials for
oxidation of the cation radicals to be less than the standard
electrode potentials for these processes (E_2's), the observed
values do permit the calculation of an upper limit of the disproportionation constant, K_{dis}. For the DPA and TH systems, these
limiting values are essentially equal.

It is of interest, then, that the $DPA^{+\cdot}/H_2O$ system reacts <u>via</u> the HRM whereas the $TH^{+\cdot}/H_2O$ system has been reported to follow the disproportionation pathway. This difference in reaction demeanor is attributable more to the electronic character of the heteroatom <u>per se</u> than to its impact on the orbital energies of the molecule. This is supported by the formation of the sulfoxide wherein the S atom has formally released two electrons and ultimately becomes a 10-electron center. Formation of such products is contingent upon charge relief <u>via</u> proton release which in this case occurs through loss of protons originally bourne by the nucleophile. It is also known that even partial charge relief imparts a measure of stability to similar addition products (3-5,7,8).

Deprotonation is a charge relief process of <u>kinetic</u> importance in the hydrolysis of $TH^{+\cdot}$. The third order dependence of the rate of this reaction upon H_2O concentration shown in Table V has been accounted for in terms of the HRM (10) as follows:

$$TH^{+\cdot} + H_2O \overset{K}{\rightleftharpoons} TH(OH_2)^{+\cdot} \tag{16}$$

$$TH(OH_2)^{+\cdot} + H_2O \overset{K}{\rightleftharpoons} TH(OH)^{\cdot} + H_3O^{+} \tag{17}$$

$$TH(OH)^{\cdot} + TH(OH_2)^{+\cdot} \overset{k}{\longrightarrow} TH(OH)^{+} + TH + H_2O \tag{18}$$

$$TH(OH)^{+} + H_2O \overset{fast}{\longrightarrow} TH(O) + H_3O^{+} \tag{19}$$

This reaction pathway involves both reversible formation and reversible deprotonation of the cation radical-nucleophile adduct prior to rate determining electron transfer with a <u>second adducted</u> radical ion [equation (18)]. The observed third order dependence of reaction rate on H_2O concentration points to the role of $TH(OH_2)^{+\cdot}$ as the oxidant <u>rather</u> than $TH^{+\cdot}$ which suggests a favorable adduction equilibrium [equation (16)]. Moreover, the predominant oxidation of $TH(OH)^{\cdot}$ by $TH^{+\cdot}$ would give rise to a rate law which is second order with respect to the concentration of H_2O.

While deprotonation is clearly involved in the formation of the hydrolysis products of $DPA^{+\cdot}$, it is of no apparent consequence to the kinetics of radical ion consumption. In this reaction scheme, the observed first

$$DPA^{+\cdot} + H_2O \xrightarrow{k} DPA(OH_2)^{+\cdot} \tag{20}$$

$$DPA(OH_2)^{+\cdot} + H_2O \longrightarrow DPA(OH)^{\cdot} + H_3O^+ \tag{21}$$

$$DPA(OH)^{\cdot} + DPA^{+\cdot} \longrightarrow DPA(OH)^+ + DPA \tag{22}$$

$$DPA(OH)^+ + H_2O \longrightarrow DPA(OH)(OH_2)^+ \tag{23}$$

$$DPA(OH)(OH_2)^+ + H_2O \longrightarrow DPA(OH)_2 + H_3O^+ \tag{24}$$

order dependencies of the reaction rate on the concentrations of $DPA^{+\cdot}$ and H_2O shown in Table V clearly indicate the absence of a deprotonation process prior to the rate determining cation radical-nucleophile encounter and points to the oxidation of $DPA(OH)^{\cdot}$ by a non-adducted cation radical [equation (22)].

The necessity of charge relief in reactions of S-centered cation radicals has been further demonstrated in the reactions of $TH^{+\cdot}$ and $PH^{+\cdot}$ with Py. Pyridination of $DPA^{+\cdot}$ gives rise to a stable dicationic addition product, $DPA(Py)_2^{++}$ (33). Pyridinations of $TH^{+\cdot}$ and $PH^{+\cdot}$, however, afford the monocationic ring substitution products, $TH(Py)^+$ and $P(Py)^+$, in which partial charge relief has been attained by proton loss from the site of substitution (6,11). The mechanism of formation of these substitution products involves initial attack by Py at an S atom, the site of highest positive charge density in the radical ion. This N-S bonded adduct is then oxidized by a second cation radical as shown in Figure 2. Subsequent nucleophilic attack by Py at a ring site on this oxidized adduct affords the observed ring-substituted product (11).

The presence of trace quantities of H_2O in these latter systems gives rise to formation of substantial quantities of the sulfoxides (Table IV). Although H_2O is inherently a less reactive nucleophile and is present in much lower concentration than Py, the formation of the sulfoxides is preferred on the basis of product stability achieved through charge relief.

In a solvent system chosen to minimize the activity of H_2O (4% (v/v) TFAn in CH_3CN), the kinetics of $TH^{+\cdot}$ consumption were found to be dependent upon the concentration of Py (Table VI) even though the reaction product is exclusively $TH(O)$ (Table IV). These results indicate rapid hydrolysis of the oxidized N-S bonded cation radical-Py adduct formed in the rate determining electron transfer step as shown in Figure 3 (11).

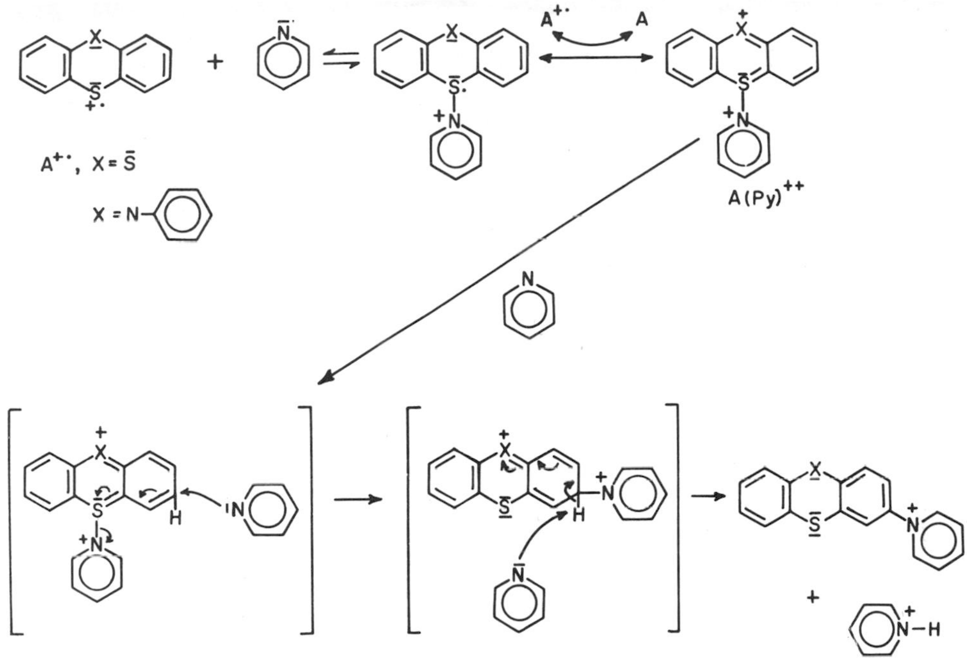

Figure 2. Proposed mechanism for pyridination of TH$^{+\cdot}$ and PH$^{+\cdot}$

The reactivity of TH$^{+\cdot}$ toward Py precluded quantitation of
the kinetics of the process giving rise to the pyridinated product.
The less reactive PH$^{+\cdot}$/Py system, however, permits the kinetic
characterization of the reaction which affords the analogous
pyridinated product. The diminished reactivity of this system is
attributable to the presence of the N heteroatom and phenyl substitution at this site precludes the manifold reactions available
to the unsubstituted phenothiazine cation radical (6,15).

For the pyridination of PH$^{+\cdot}$, the rate of disappearance of
radical ion was found to be second order with respect to cation
radical concentration, first order with respect to nucleophile
concentration and inversely dependent upon concentration of
precursor as shown in Table VI. This kinetic demeanor strongly
suggests the involvement of disproportionation en route to product
formation. However, careful analysis of the kinetic data, together
with a reliable measurement of K_{dis} for the PH system, dictates the
dismissal of the disproportionation mechanism as the dominant
reaction route in favor of the HRM in which nucleophilic attack by
Py on the N-S bonded cation radical adduct is involved in the rate
determining process (11).

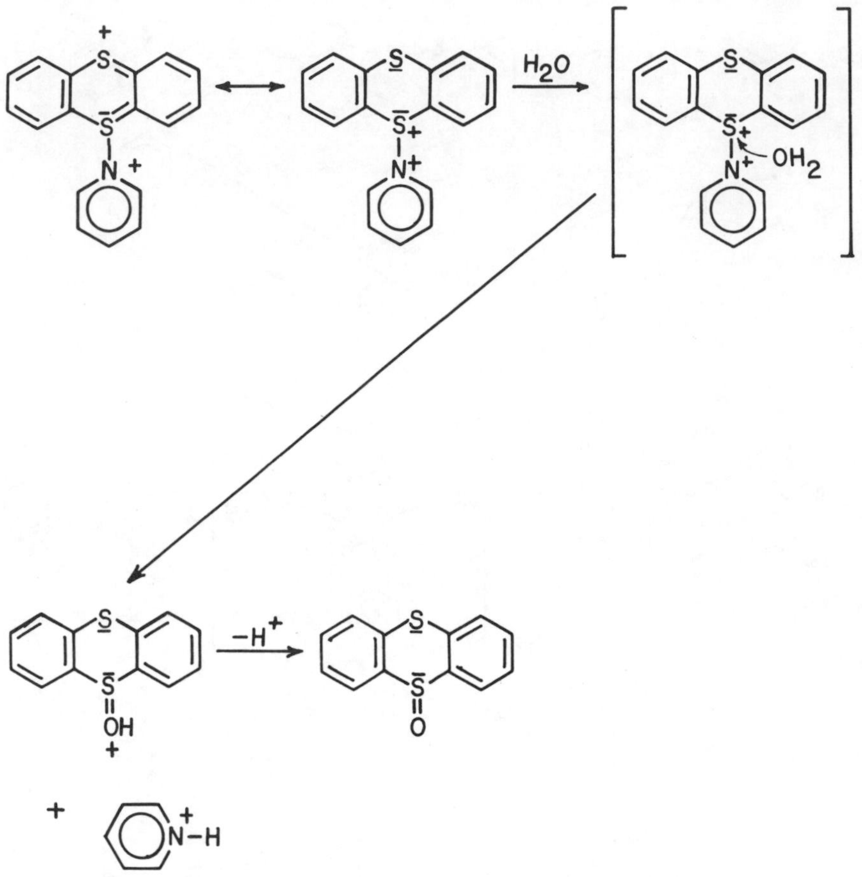

Figure 3. Proposed mechanism for pyridine catalyzed formation of TH(O) from $TH^{+\cdot}$ in acetonitrile.

4.   CONCLUSIONS

Work conducted in these laboratories (10,11,31,33-35,70) indicates the applicability of the HRM to a wide variety of seemingly divergent reaction systems. Kinetic control at different junctures within the framework of this mechanism gives rise to rate laws which are (a) first order in both cation radical and nucleophile concentrations (31,35,70); (b) second order in cation radical concentration and first order in nucleophile concentration (10,34); and (c) second order in cation radical concentration, first order in nucleophile concentration, and inversely dependent upon precursor concentration (11). The first of these rate laws corresponds to kinetic control by the formation of the adducted cation radical followed by oxidation of this adduct by a second cation

radical.   The second rate law reflects rapid, reversible formation
of the adducted cation radical followed by <u>rate determining</u>
<u>oxidation</u> of this species by a second cation radical.   The third
rate law reflects not only reversible formation and oxidation of
the cation radical-nucleophile adduct, but also <u>involvement of</u>
<u>this</u> <u>oxidized</u> <u>form</u> <u>in</u> <u>the</u> <u>rate</u> determining <u>process</u>.   Distinction
between disproportionation and the HRM for systems which react
according to this third rate law is only possible when the value
of K_{dis} is known to be sufficiently small such that absolute rate
theory cannot account for the observed system dynamics by this
pathway (11).

In systems where the adducted cation radical is reversibly
formed and capable of undergoing charge relief <u>via</u> proton release,
it has been shown that the acidity of these adducts is of major
consequence to the observed reaction kinetics (10,11).

On the basis of the kinetics observed for the reactions of the
C-centered cation radical $DPA^{+\cdot}$ with nine different protic, aprotic
and anionic nucleophiles which afford both addition and electron
transfer products, a scale of relative nucleophilicities has been
established (70).   These parameters have been shown to correlate
extremely well with those determined for these same species (64,
69,70) for nucleophilic displacement reactions at tetrahedral
carbon.   This correlation is to be expected if, indeed, a common
transition state is achieved in the rate determining steps of <u>both</u>
addition <u>and</u> electron transfer reactions.

5.   SUMMARY

Stopped-flow and various spectroelectrochemical techniques
have been employed to delineate the kinetic and mechanistic details
of the reactions of protic and aprotic nucleophiles with the cation
radicals derived from 9,10-diphenylanthracene, thianthrene, and
10-phenylphenothiazine.   A scale of relative nucleophilic reactivities for various protic, aprotic, and anionic nucleophiles has
been established for both cases in which addition products are
noted as well as those in which the nucleophile serves to reduce
the cation radical to DPA.   These relative reactivities toward
$DPA^{+\cdot}$ in acetonitrile have been shown to correlate extremely well
with the classical nucleophilicities of these electron donors.

Although the oxidative electrochemistry of DPA, TH and PH and
their respective cation radicals is quite similar, the presence of
N and S heteroatoms in the latter two systems has been shown to
markedly affect the reactivities of these cation radicals toward
the same nucleophiles.   A general half-regeneration mechanism has
been shown to accommodate all cation radical/nucleophile reactions

studied in which addition or substitution products are observed.
Cation radical/nucleophile adduct formation and its effect on
kinetic control of the reaction is discussed within the framework
of the half-regeneration scheme.

6.   REFERENCES

(1)   Y. Murata and H. J. Shine, J. Org. Chem., 34, 3368 (1969).

(2)   H. J. Shine and Y. Murata, J. Am. Chem. Soc., 91, 1872
      (1969).

(3)   J. J. Silber and H. J. Shine, J. Org. Chem., 36, 2923 (1971).

(4)   J. J. Silber, Ph.D. Thesis, Texas Tech University, Lubbock,
      Texas, 1972.

(5)   H. J. Shine and J. J. Silber, J. Am. Chem. Soc., 94, 1026.
      (1972).

(6)   H. J. Shine, J. J. Silber, R. J. Bussey and T. Okuyama,
      J. Org. Chem., 37, 2691 (1972).

(7)   H. J. Shine and K. Kim, Tetrahedron Lett., 99 (1974).

(8)   K. Kim and H. J. Shine, Tetrahedron Lett., 4413 (1974).

(9)   U. Svanholm, O. Hammerich and V. D. Parker, J. Am. Chem.
      Soc., 97, 101 (1975).

(10) J. F. Evans and H. N. Blount, J. Org. Chem., 42, 976 (1977).

(11) J. F. Evans, J. R. Lenhard and H. N. Blount, J. Org. Chem.,
      42, 983 (1977).

(12) C. Barry, G. Cauquis and M. Maurey, Bull. Soc. Chim. Fr.,
      2510 (1966).

(13) S. R. Mani and H. J. Shine, J. Org. Chem., 40, 2756 (1975).

(14) T. N. Tozer and L. Dallas Tuck, J. Pharm. Sci., 54, 1169
      (1965).

(15) P. Hanson and R. O. C. Norman, J. Chem. Soc., Perkin II, 264
      (1973).

(16) B. K. Bandlish, A. G. Padilla and H. J. Shine, J. Org. Chem.,
      40, 2590 (1975).

(17) G. Cauquis and M. Maurey-Mey, Bull. Soc. Chim. Fr., 3588
      (1972).

(18) H. J. Shine and L. R. Shade, J. Heterocycl. Chem., 11, 139
      (1974).

(19) R. E. Sioda, J. Phys. Chem., 72, 2322 (1968).

(20) L. Papouchado, R. N. Adams and S. W. Feldberg, J Electroanal.
      Chem., 21, 408 (1969).

(21) C. R. Manning, Ph.D. Thesis, University of Kansas, Lawrence,
      Kansas, 1969.

(22) G. Manning, V. D. Parker and R. N. Adams, J. Am. Chem. Soc.,
      91, 4584 (1969).

(23) G. C. Grant and T. Kuwana, J. Electroanal. Chem., 24, 11
      (1970).

(24) H. N. Blount and T. Kuwana, J. Electroanal. Chem., 27, 464
      (1970).

(25) P. T. Kissinger and C. N. Reilley, Anal. Chem., 42, 12
      (1970).

(26) L. Marcoux, J. Am. Chem. Soc., 93, 537 (1971).

(27) G. Mengoli and G. Vidotto, Makromol. Chem., 150, 277 (1971).

(28) V. D. Parker, J. Electroanal. Chem., 36, App. 8 (1972).

(29) R. P. Van Duyne and C. N. Reilley, Anal. Chem., 44, 158
      (1972).

(30) C. J. Ludman, E. M. McCarron and R. F. O'Malley, J. Electro-
      chem. Soc., 119, 874 (1972).

(31) H. N. Blount, J. Electroanal. Chem., 42, 271 (1973).

(32) U. Svanholm and V. D. Parker, Acta Chem. Scand., 27, 1454
      (1973).

(33) D. T. Shang and H. N. Blount, J. Electroanal. Chem., 54, 305
      (1974).

(34) J. F. Evans and H. N. Blount, J. Org. Chem., 41, 516 (1976).

(35) J. F. Evans and H. N. Blount, J. Phys. Chem., 80, 1011
      (1976).

(36) L. Eberson and H. Schafer in "Fortschritte der chemischen Forschung," Vol. 21, A. Davison, et al., Eds., Springer-Verlag, Berlin, 1971.

(37) M. M. Baizer, Ed., "Organic Electrochemistry," Marcel Dekker, Inc., New York, 1973.

(38) N. L. Weinberg in "Techniques of Chemistry," A. Weissberger, Ed., Vol. V, Part I, John Wiley and Sons, New York, 1974.

(39) N. L. Weinberg in "Techniques of Chemistry," A. Weissberger, Ed., Vol. V, Part II, John Wiley and Sons, New York, 1975.

(40) L. Eberson and K. Nyberg in "Advances in Physical Organic Chemistry," V. Gold, Ed., Vol. 12, Academic Press, New York, 1976.

(41) A. J. Bard, A. Ledwith and H. J. Shine in "Advances in Physical Organic Chemistry," V. Gold, Ed., Vol. 13, Academic Press, New York, 1976.

(42) K. W. Fung, J. Q. Chambers and G. Mamantov, J. Electroanal. Chem., 47, 81 (1973).

(43) O. Hammerich and V. D. Parker, Electrochim. Acta, 18, 537 (1973).

(44) J. W. Strojek, T. Kuwana and S. W. Feldberg, J. Am. Chem. Soc., 90, 1353 (1968).

(45) N. Winograd and T. Kuwana, J. Am. Chem. Soc., 92, 224 (1970).

(46) N. Winograd and T. Kuwana, J. Am. Chem. Soc., 93, 4343 (1971).

(47) R. G. Pearson, Ed., "Hard and Soft Acids and Bases," Dowden, Hutchinson and Ross, Inc., Stroudsburg, PA, 1973.

(48) J. O. Edwards, J. Am. Chem. Soc., 76, 1540 (1954).

(49) J. O. Edwards and R. G. Pearson, J. Am. Chem. Soc., 84, 16 (1962).

(50) L. Eberson, J. Chem. Soc. Chem. Comm., 826 (1975).

(51) G. Cauquis and G. Pierre, Bull. Soc. Chim. Fr., 736 (1976).

(52) U. Svanholm and V. D. Parker, J. Am. Chem. Soc., 98, 997 (1976).

(53) J. F. Evans, H. N. Blount and C. R. Ginnard, J. Electroanal.
      Chem., 59, 169 (1975).

(54) J. F. Evans and H. N. Blount, Anal. Lett., 7, 445 (1974).

(55) S. Andreades and E. Zahnow, J. Am. Chem. Soc., 91, 4181
      (1969).

(56) O. W. Webster, W. Mahler and R. E. Benson, J. Am. Chem.
      Soc., 84, 3678 (1962).

(57) E. A. Deardorff, Ph.D. Dissertation, University of Delaware,
      Newark, Delaware, 1970.

(58) T. Kuwana and N. Winograd in "Electroanalytical Chemistry,"
      Vol. 7, A. J. Bard, Ed., Marcel Dekker, Inc., New York, 1974.

(59) W. von Benken and T. Kuwana, Anal. Chem., 42, 1114 (1970).

(60) T. Kuwana and W. R. Heineman, Acc. Chem. Res., 9, 241 (1976).

(61) A. A. Pilla, J. Electrochem. Soc., 118, 702 (1971).

(62) N. Winograd, H. N. Blount and T. Kuwana, J. Phys. Chem., 73,
      3456 (1969) and references therein.

(63) Estimated from the aqueous data of W. M. Latimer, "The
      Oxidation States of the Elements and Their Potentials in
      Aqueous Solutions," 2nd ed., Prentice-Hall, Inc., Englewood
      Cliffs, New Jersey, 1952.

(64) R. G. Pearson, H. Sobel and J. Songstad, J. Am. Chem. Soc.,
      90, 319 (1968).

(65) Estimated from the data of G. Klopman, J. Am. Chem. Soc., 90,
      223 (1968).

(66) From the data of ref. 49, corrected to methanol.

(67) From the data of ref. 68, corrected to methanol.

(68) A. Fischer, W. J. Galloway and J. Vaughan, J. Chem. Soc.,
      3596 (1964).

(69) C. G. Swain and C. B. Scott, J. Am. Chem. Soc., 75, 141
      (1953).

(70) J. F. Evans and H. N. Blount, submitted for publication.

(71) L. S. Marcoux, R. N. Adams and S. W. Feldberg, J. Phys. Chem.,
      73, 2611 (1969).

MEASUREMENT OF CONFORMATIONAL RATES AND EQUILIBRIA BY CYCLIC VOLTAMMETRY

Dennis H. Evans and Stephen F. Nelsen

Department of Chemistry, University of Wisconsin

Madison, Wisconsin 53706

1. INTRODUCTION

Throughout the history of electrochemistry there has been keen interest in the details of the most central and fundamental step in any electrode reaction, the heterogeneous electron transfer reaction. Two aspects of this problem are configurational in nature. (1) Is there a particular nuclear configuration of the reactant molecule which is most favorable for electron transfer? (2) Is there a particular orientation of the reactant molecule with respect to the electrode surface which is most favorable for electron transfer? There is very little direct experimental evidence bearing on either of these questions, though the literature is replete with speculation.

In this chapter we will review what we believe to be the first evidence that a particular conformation of a reactant molecule is more readily oxidized than its other principal conformation, i.e., we will directly address the first of the above questions.

The reactants we will discuss are tetraalkylhydrazines. The conformations of tetraalkylhydrazines have been recently studied using photoelectron spectroscopy (1) and low temperature ^{13}C-NMR (2), and the radical cations derived by one electron oxidation using ESR (3). Cyclic voltammetry has proven to be a useful tool for characterization of the redox properties of tetraalkylhydrazines, (4) and it was in the course of such studies that it was discovered that certain of these compounds show two oxidation peaks at low temperatures and fast scan rates (5).

2.   THEORETICAL CONSIDERATIONS

Before discussing the data for tetraalkylhydrazines, it may
be useful to consider in more general terms the possible effects
of molecular conformation on electrochemical behavior and to review some of the earlier electrochemical studies which relate to
the question at hand.

As a framework for the discussion, let us consider the substituted ethanes shown in Figure 1. Here the Y and Z substituents
are either *anti* or *gauche* with respect to each other. The interconversion of the *anti* and *gauche* conformations is characterized
by rate constants (k_{ag} and k_{ga}) and an equilibrium constant, K.
Each conformation may be reduced in an electrode reaction with a
characteristic standard potential, standard heterogeneous electron
transfer rate constant and electron transfer coefficient. The
products (here shown as *anti* and *gauche* radical anions) will interconvert with characteristic rate constants (k_{ag}^{-} and k_{ga}^{-}) and each
may undergo additional chemical reactions.

What conditions are necessary for the *anti* and *gauche* reactants
to show separate reduction processes as detected by voltammetry?

First, it is essential that the rate of interconversion of
reactant conformations be slow on the voltammetric time scale.
Otherwise, all of the reactant molecules would be reduced at the
potential of the more easily reduced conformation. In practice,
this will require an activation energy greater than about 8 kcal/
mole in order to see separate reduction processes at easily attainable temperatures and scan rates.

The second requirement is that the electrochemistry of each
conformation be sufficiently different to cause reduction to occur
at potentials different enough to be resolved by voltammetry.
There are two types of factors which can cause the difference:

(1) Thermodynamic factors: These appear in the standard
potentials, E_a° and E_g°. The relative free energies of the four
species in the scheme govern these standard potentials.

(2) Kinetic factors: There are two types. In the first, the
standard heterogeneous electron transfer rate constants for the
reduction of the two conformations, $k_{s,a}$ and $k_{s,g}$, can differ,
causing one to be reduced at a different potential than the other.
In other words, the activation overpotentials differ. In the
second type, which is probably less common, the rate constants of
the chemical reactions of the product conformations, k_a' and k_g',
differ sufficiently to shift the reduction potentials of the
reactant conformations to different values. In principle, chemical

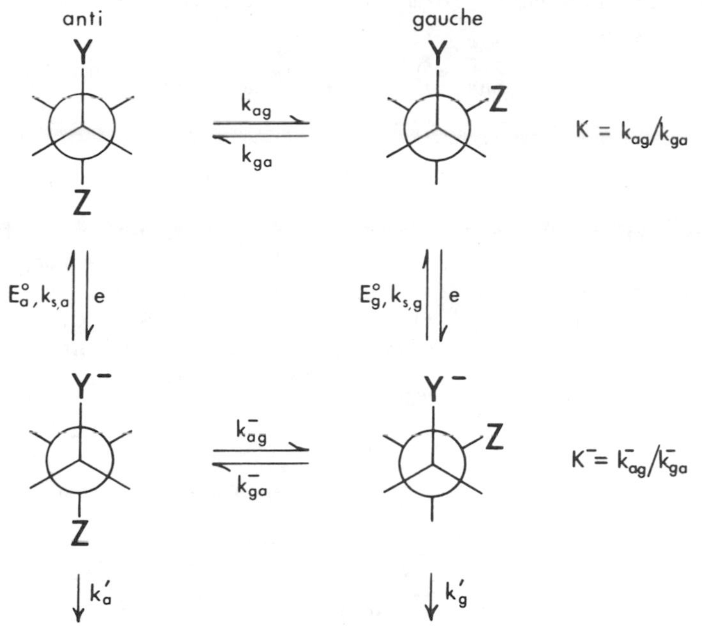

Figure 1. Conformation and Redox Scheme for a Disubstituted Ethane.

reactions of the reactant conformations prior to electron transfer could have a similar effect.

It should be emphasized that in order for these thermodynamic and kinetic factors to cause two voltammetric reduction peaks to be observed, the rates of conformational interconversion must be slow. In the limit of perfectly mobile conformational equilibria and all of the reduction reversibly proceeding via the $anti$ conformer, the voltammetric experiments will detect a single reduction process with an apparent standard potential,

$$E^{\circ}_{obs} = E^{\circ}_{a} + \frac{RT}{F} \ln \frac{(K^{-} + 1)}{(K+1)} \tag{1}$$

Similarly, if the *gauche* form is the reactive conformer

$$E^{\circ}_{obs} = E^{\circ}_{g} + \frac{RT}{F} \ln \frac{K(K^{-} + 1)}{K^{-}(K+1)} \tag{2}$$

3.   REVIEW OF EARLIER WORK

There appear to be no examples of systems where different standard potentials have been measured for conformations differing by angle of rotation about a single bond.   Data are available, however, for reductions of the geometrical isomers of various olefins (Table I).   In these systems, of course, the interconversion of reactant isomers is very slow though the radical anion products may isomerize rapidly.

In general, the *cis* isomers are somewhat more difficult to reduce, a fact attributed to greater charge repulsion by the

TABLE I

Difference in Standard Potentials for Geometrical Isomers
of Olefins

Pair	$E^{\circ}_{trans} - E^{\circ}_{cis}, mV^a$	Reference
cis- and *trans*-stilbene	~30	6
diethyl maleate, diethyl fumarate	280^b	7
cis- and *trans*-1,4-diphenyl-1,4-butenedione	~300	7
cis- and *trans*-thioindigo	c	8

[a]Potentials refer to olefin/anion radical couples in DMF.
[b]Corrected for effects of dimerization of radical anions.
[c]Too small to detect.

cis-groups in the radical anion.   This also causes the cis-radical
anions to isomerize to the trans forms.   For example, the cyclic
voltammetric reduction of diethyl maleate shows a one electron
reduction peak with no associated anodic peak because the cis-
radical anion rapidly isomerizes to the trans-radical anion which
has a more positive oxidation potential.

An interesting variation of this scheme has been discovered by
Rieke, Kojima and Öfele (9), who studied several dicarbene metal
carbonyl complexes such as 1.   In this case the cis isomer is the
more difficult to oxidize ($\Delta E^{\circ} \sim 0.2V$) and it is also the stabler
form.   The cations isomerize quickly so that attempted oxidation of
trans-1 at potentials negative of E°_{cis}, results in redox catalyzed
isomerization in a reaction layer near the electrode and almost no
net current.   This reaction scheme has been called an ECE process
(10).

trans-1   cis-1

$$trans\text{-}1 \xrightarrow{-e} (trans\text{-}1)^{+} \longrightarrow (cis\text{-}1)^{+} \xrightarrow{+e} cis\text{-}1$$

Biathrone (2) is another molecule whose isomers show different electrochemical behavior.   In this case, the isomers are
thermochromic forms of 2, a low temperature (yellow) form and a
high temperature (green) form.   There is some controversy about
the structure of the two forms but the preponderant evidence seems
to indicate that the yellow form is fairly flat while the green
form is twisted about the central bond so that the two anthrone
systems are perpendicular.   In any case, the green form is much
easier to reduce and a small kinetic current is observed at
positive potentials when voltammetry is performed on solutions of
the yellow form.   This current has been interpreted as being due
to the conversion yellow → green occurring in a reaction layer near
the electrode.   Thus bianthrone is representative of a class of

2

molecules where different isomers have distinctly different standard potentials and the interconversion of isomers is slow but
detectable at temperatures near ambient.

Returning to molecules in which conformational change is
effected by rotation about single bonds, one finds no cases of
purely thermodynamic factors causing different electrochemical behavior for different conformations. However, the work of Závada,
Krupicka and Sicher (15) suggests that conformers of cyclic
vic-dibromides are reduced at distinctly different potentials and
that the differences are due to differing rates of the electrode
reactions.

These authors studied a series of twenty-one *vic*-dibromides
many of which were incorporated in rigid cyclic systems so that the
dihedral angle between the two carbon-bromine bonds was known. They
found that dihedral angles near 180° were most favorable for reduction as indicated by half-wave potentials near -0.8 V vs. SCE. On
the other hand, dihedral angles of 60-120° were unfavorable for
reduction with half-wave potentials of approximately -1.6 V. Angles
near 0° also were associated with less negative half-wave potentials. The data for compounds 3-5 illustrate the trend. Thus the
dihedral angle appears to be the most important parameter governing
the half-wave potentials. It is assumed that the differences in
potential are due to differences in rate of the irreversible
electrode reactions with dihedral angles near 180° or 0° associated
with large rate constants.

The rigid *trans*-dibromide 5 cannot convert to the diaxial form
due to the bulk of the *tert*-butyl group. Hence, it cannot achieve
a dihedral angle of ~180° which would probably permit reduction at
less negative potentials. This was verified by Casanova and Rogers
(16) who subjected 5 to electrolysis at -0.86 and -1.30 V vs. SCE
and found that no reduction occurred. *trans*-Dibromides which are
not locked in the diequatorial conformation are easily reduced
because any fraction existing in the diequatorial form can rapidly
convert to the diaxial conformation and be reduced. For example,

the half-wave potential of *trans*-1,2-dibromocyclohexane is -1.04 V (15).

3

$E_{1/2}$ = -0.86 V vs. SCE

4

-1.21 V vs. SCE

5

$E_{1/2}$ = -1.67 V vs. SCE

Though the data for the *vic*-dibromides suggest that different conformations will have distinctly different reduction potentials, all of the conclusions are based on studies of compounds of rigid conformation and no case is yet known where two conformations of the same molecule have been shown to be reduced at different potentials. To see such an effect it will be necessary to have relatively slow rates of conformational change, i.e., large activation energies and sub-ambient temperatures.

4. CYCLIC VOLTAMMETRIC STUDIES OF TETRAALKYLHYDRAZINES

Most tetraalkylhydrazines are electrochemically oxidized to rather persistent radical cations (radical lifetimes at millimolar concentration are often minutes to hours) (4b). Cyclic voltammetry (cv) at slow scan rates shows a nearly reversible one electron oxidation process with an associated reduction peak for the radical cation. It was discovered (5) that at low temperatures some hexahydro- and tetrahydro- pyridazines show two oxidation peaks but only one reduction peak. We shall first present the qualitative

arguments for the peak assignments and then review the results of
a quantitative analysis of the cv data in terms of conformational
rates and equilibria.

Qualitative Considerations

A description of the appearance of the cv curves obtained for
1,2,3-trimethylhexahydropyridazine (6) illustrates our observations. Although 6 shows an ordinary~cv curve at room temperature,
different behavior is observed at -55° in butyronitrile (17). As
shown in Figure 2, at a 50 mV/sec sweep rate, some extra current
is observed at higher potential than the room temperature oxidation peak. As the sweep rate is increased, this extra current
grows into a second oxidation peak, which moves progressively to
higher potential, and the first peak decreases in size relative to
the second. These effects were completely physically reversible,
the second peak disappearing when

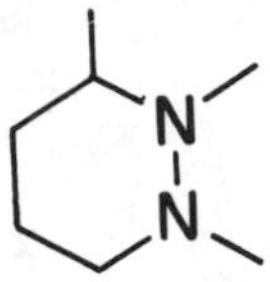

6

the temperature was raised, and reappearing again upon lowering the
temperature. Qualitatively similar behavior was seen at a platinum
electrode and in other solvents (acetone, methylene chloride).

Two oxidizable species which interconvert rapidly at high
temperatures but more slowly at low temperatures are clearly required to explain the cv curves observed. In view of the basicity
of the hydrazines, it might be suggested that the first oxidation
peak could be caused by oxidation of the free hydrazine, and the
second by oxidation of the protonated form, present because of
acidic impurities. The following CEE scheme (1) might be argued
to be a possibility (B is the hydrazine).

Scheme 1 $\qquad\qquad BH^{+} \underset{k_b}{\overset{k_f}{\rightleftharpoons}} B + H^{+}$

$$B \xrightarrow{E_1^o} B^{+\cdot} + e \text{ (reversible)}$$

$$BH^{+} \xrightarrow{E_2^{o'}} B^{+\cdot} + H^{+} + e \text{ (irreversible)}$$

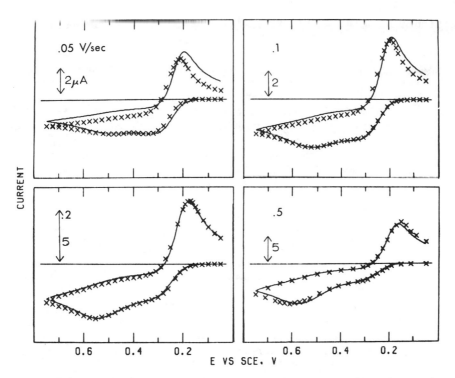

Figure 2.   Plots of experimental (X) and calculated (solid line) cyclic voltammograms for 6 at -55°, scan rate 50, 100, 200, 500 mV/sec.   The following parameters were set as constants for these simulations: Apparent E_{ee}° = 0.255 V, $K_{eq}\sqrt{k_t}$ = 0.68 sec^{-1}, α_1 = 0.5, $k_{s,1}/\sqrt{D}$ = 20.2 $sec^{-\frac{1}{2}}$, α_2 = 0.7, $k_{s,2}/\sqrt{D}$ = 4.34 x 10^{-3} $sec^{-\frac{1}{2}}$, referred to arbitrary $E_{ae}^{\circ\prime}$ = +0.125 V (5b).

The fact that voltammograms qualitatively similar to those obtained
in butyronitrile were observed in dichloromethane and acetone
argues against Scheme 1. This scheme was effectively eliminated
when it was found that addition of even a 50-fold excess of butyric
acid did not alter the relative peak heights for 6 appreciably, as
is required by this scheme (5b).

We suggested instead, that the two species showing separate
electrochemistry at low temperature were different conformations
of 6 (5a). N,N'-Dialkylhexahydropyridazines exist in conformations
with diequatorial alkyl groups (H_{ee}) and those with axial, equatorial alkyl groups (H_{ae}). If the equilibrium between H_{ae} and H_{ee}
were slow compared to the sweep rate, different oxidation peaks
would be observed for oxidation of the two conformers if their peak
potentials were sufficiently different. For 6, photoelectron
spectroscopy (1b,1e) had previously established that H_{ae} conformations predominate over H_{ee} ones, so the second oxidation peak,
which is larger than the first at low temperature and fast scan
rate, was assigned to the H_{ae} conformations.

H_{ee} H_{ae}

If the assignment of the two oxidation peaks to H_{ee} and H_{ae}
conformations is correct, one should observe a larger second oxidation peak than the first whenever axial, equatorial conformations
predominate. The H_{ae} conformations of 7-9 were shown by ^{13}C-NMR
to predominate (2), and all three have second oxidation peaks
which are larger than the first in low temperature cv experiments
(5b).

Significantly, photoelectron spectroscopy (1b, 1e) and
^{13}C-NMR spectroscopy show that 10 and 11 exist exclusively in H_{ee}
conformations, and neither showed any trace of a second oxidation
process under any conditions (5). Quite low temperature (-70°) was
required to see the appearance of a second oxidation peak for 12,
and the second peak remained

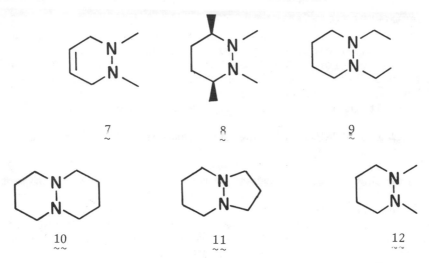

smaller than the first at the fastest scan rate used (1 V/sec).
Low temperature NMR work showed that the H_{ee} conformation slightly
predominates for 12 (2a, 2c). The temperature required for observation of a second oxidation peak will depend both upon the size
of the equilibrium constant (a small value of $K_{eq} = [H_{ee}]/[H_{ae}]$ will
allow the second peak to be observed at a higher temperature) and
the activation energy for $H_{ae} \rightarrow H_{ee}$ (a large activation energy will
allow the second peak to be observed at a higher temperature). A
lower value of K_{eq} must be responsible for being able to see the
second peak for 7-9 at higher temperature than for 12, since the
activation energy for $H_{ae} \rightarrow H_{ee}$ conversion for 7-9 is expected to
be equal or lower than that of 12 (2). The activation energy
effect is clearly shown by 13, which has almost the same K_{eq} as 12
but shows the second cv oxidation peak at temperatures as high as
0°C (18). 13 has a significantly larger activation energy for
$H_{ae} \rightarrow H_{ee}$ conformational process (2c) than does 12.

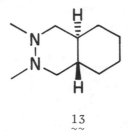

13

Only one wave for reduction of the hydrazine radical cation
to the neutral hydrazine was observed for any compound. This was
interpreted as follows (5b). Earlier ESR work (3) had shown that
the radical cations have quite different conformations from those

of the neutral hydrazines, because the lone pair orbitals have a strong restoring force towards keeping the lone pair orbitals coplanar. The radical cations are flattened at nitrogen, but have low activation energy for double nitrogen inversion (3c). One would expect, then, that the conformations of the cation radicals would be rapidly interconverting at accessible temperatures, which would result in only one reduction peak being observed.

From the observation that the peak potential of the second oxidation peak moved to higher potential considerably more rapidly than did that of the first peak as the scan rate was increased, it was clear that electron transfer from H_{ae} must be slower than that from H_{ee}. The electron transfer giving rise to the second peak is slow, and appears electrochemically totally irreversible. In contrast, the H_{ee} oxidation (the first peak) appears to be nearly electrochemically reversible. It was suggested that the large difference in lone pair-lone pair dihedral angle, about $60°$ for H_{ae}, but $180°$ for H_{ee}, so that the lone pair orbitals are approximately coplanar, as required in the cation radical, is responsible for the difference in the rates of electron transfer (5b).

The minimal kinetic scheme which arose from the above qualitative considerations is that shown in Scheme 2, in which $k(H_{ee} \rightarrow H_{ae})$ and $k(H_{ae} \rightarrow H_{ee})$ are first order rate constants, and $E°$, k_s and α are the standard potential, standard heterogeneous electron transfer rate constant, and electron transfer coefficient for the respective electrode reactions.

Generation of Theoretical Voltammograms by Digital Simulation

To test whether the observed voltammograms would actually correspond to the above simple scheme, in which the conformational interconversion between H_{ee} and H_{ae} is completely unaffected by the presence of the electrode, the method of digital simulation was used to generate the cv curves which would result if Scheme 2 operated.

$$H_{ee} \underset{k(H_{ae} \rightarrow H_{ee})}{\overset{k(H_{ee} \rightarrow H_{ae})}{\rightleftharpoons}} H_{ae}$$

$$E°_{ee}, k_{s,1}, \alpha_1 \quad \searrow -e \qquad -e \swarrow \quad E°'_{ae}, k_{s,2}, \alpha_2$$

$$H^{+}_{\cdot}$$

Scheme 2

Theory for cyclic voltammetry for this CEE reaction scheme had been presented earlier by Shuman and Shain (19) who restricted their treatment to the case of reversible electron transfer for the first peak. The digital simulation technique (20) modified by using the steady state approximation (21) was chosen because of the ease with which electron transfer kinetics (quasireversibility) could be introduced into the model.

Though Ružic̆ and Feldberg (21) demonstrated that their "heterogeneous equivalent" approach gave accurate simulations for the CE mechanism for chronoamperometry, and one of us (22) has shown that equally satisfactory results are obtained for the EC and EC$_{Dimer}$ mechanisms for cyclic voltammetry, it was deemed necessary to check the veracity of the approach for the CE process for cyclic voltammetry.

This was done by comparing results of the digital simulations using the "heterogeneous equivalent" approach (21) (our computer program will be provided upon request) with the results for the CE case obtained by the well established integral equation procedure (23,24). Specifically, the digital simulations were performed using a total of 2080 time units for the entire cycle. The values of the current function so obtained were within ±0.001 (average deviations; largest deviation was 0.009) of the earlier results (uncertainty of these is +0.000, -0.002 (23)) for values of the kinetic parameter, $\kappa = a^{1/2}/k_t^{1/2}K_{eq}$, ranging from 0 to 100 ($k_t = k_f + k_b$ and $a = nFv/RT$ where v is the scan rate). Thus the procedure produced theoretical voltammograms of sufficient accuracy for comparison with the experimental data.

The steady state approximation should and does (22) fail when a/k_t becomes very large (e.g., fast experiments and/or small rate constants). In addition, the calculations of Nicholson and Shain (23) are not accurate for large a/k_t due to a restrictive assumption in their numerical integrations. Hence, it seemed possible that the good agreement cited above for values of κ as large as 100 could have resulted from approximately similar failure of both types of calculation. However, direct simulation (20) for $\kappa = 100$ (no steady state assumption) produced values for the current function agreeing to within 0.001 unit with each of the other approaches. Thus it was concluded that the steady state approximation gives accurate results for values of the kinetic parameter up to 100 (5b). Larger values are not normally of experimental interest.

The accuracy of digital simulations for generating cyclic voltammograms with rate-limiting electron transfer had been verified in earlier work (25).

Comparison of Experimental CV Curves and Digital Simulations

It was found that a uniform method of electrode surface
preparation was critical for obtaining reproducible results. The
position of the second oxidation peak was particularly sensitive
to all experimental variables. A freshly polished gold electrode
was held at -3.0 V vs. SCE for 3 seconds, and then at +3.0 V for
3 seconds. This pretreatment was followed by an aging, allowing
the electrode to stand at a potential where no current is passed
for several minutes. If cyclic voltammograms were run immediately
after the electrochemical pre-treatment, a second oxidation peak
was observed even at room temperature and 100 mV/sec scan rates for
some compounds, but the position of this second peak drifted
towards the first with time, and after a few minutes, only a
single, "reversible" oxidation wave was observed. Interestingly,
the extra peak was never observed for 10 or 11, which exist exclusively in diequatorial conformations. There obviously is much
to be learned about the surface of these "gold" electrodes, but
pre-treatment and aging of the electrode as described above gave
reproducible cv curves for many hours with the compounds which
were studied.

Solvent purity was also critical. The curves originally
published for 6 (5a) have a smaller reduction peak than those obtained later (Figure 2). It was found that the size of the reduction peak was especially sensitive to the quality of the butyronitrile used, and that this peak was consistently larger if the
solvent was given a final purification by a careful distillation
employing a spinning-band column.

Theoretical voltammograms which were generated are presented
in Figure 2 along with experimental data for four scan rates for
the oxidation of 1,2,3-trimethylhexahydropyridazine, 6. The
values of all simulation parameters were held constant for the
simulations shown, which were based upon linear diffusion, with
the diffusion coefficients of all species assumed equal. The
shape of the first peak and its dependence upon scan rate are
indicative of a CE process (23), and it is apparent from Figure 2
that the model adequately duplicates this feature. The second
peak becomes prominent at the more rapid scan rates, and its height
relative to the first peak as well as its shape and position are
well simulated by the simple, irreversible oxidation of H_{ae}
assumed in Scheme 2 with $\alpha_2 = 0.7$. The observed currents at
potentials positive of the second peak are somewhat greater than
the simulated values because of the onset of the oxidation of the
radical cation, a process not included in the simulations.

A single reduction peak is seen on the return half-cycle. At
slow scan rates the observed reduction peak current is less than

predicted, which may indicate a slow decomposition of the radical
cation. As mentioned above, the reduction peak current was distinctly lower when the solvent had not been distilled through a
spinning band column. Better agreement was found when such a
decomposition reaction was included in the simulations, but this
refinement was not thought to enhance significantly the analysis
of the data.

The separation between the first anodic peak and the cathodic
peak increased with increasing scan rate, an effect caused by a
combination of partially rate-limiting electron transfer and
uncompensated solution resistance. The maximum error due to the
latter factor was estimated to be about 60 mV by taking the product
of the largest current observed (20 µA) and the total cell resistance (3 kΩ at -55°) as measured from counter to working electrode.
The actual uncompensated resistance was less than 3 kΩ. The shapes
of the first oxidation peak and the reduction peak were simulated
by using partially rate-limiting electron transfer both because
the resistance effects were small, and because the effects of
uncompensated resistance and slow electron transfer are extremely
similar (26-28). The two effects were not separated so the
electron transfer kinetic parameters used in the simulation are of
very limited significance (5b).

Returning to the second oxidation peak, which was attributed
to irreversible oxidation of H_{ae}, it is of interest to identify
the factors which make possible a large enough separation between
the first and second peaks to permit their clear resolution. As
outlined earlier, the first factor is the difference between the
standard potentials ($E_{ae}^{o'} - E_{ee}^{o}$). If H_{ae} is oxidized directly to
the radical cation, $E_{ae}^{o'}$ only differs from E_{ee}^{o} by the difference in
free energy of H_{ae} and H_{ee}. Although the NMR spectrum of $\underset{\sim}{6}$ is too
complex for analysis (2d), photoelectron spectroscopy has shown
that H_{ae} conformations predominate, and that at least a few percent
of H_{ee} conformations are present in the vapor phase (1e). In all
cases where the comparison has been possible, the techniques of
^{13}C-NMR and photoelectron spectroscopy have been in excellent
agreement for conformational preferences. If there were 5% H_{ee}
conformations present at -50°, $\Delta G°$ ($\underset{\sim}{6}ee$ -$\underset{\sim}{6}ae$) would be 1.3 kcal/
mole, corresponding to only a 56 mV difference in E°, far too
small to allow resolution of the two oxidation peaks. The H_{ee} form
is 0.3 kcal/mole stabler than the H_{ae} form for $\underset{\sim}{10}$, yet two oxidation peaks were observed. One possibility would be that H_{ae} is
oxidized to a cation resembling it in geometry, so that $E_{ae}^{o'}$ is
determined not by adiabatic, but by vertical electron removal.
Even in this case (which was thought to be unlikely) the separation
of the peaks observed cannot be caused by E° differences, since
photoelectron spectroscopy shows that the vertical ionization
potentials differ by under 300 mV (1e).

Thus any difference in standard potentials for H_{ae} and H_{ee} is not sufficient to explain the resolution of the two oxidation peaks. It is necessary that the heterogeneous electron transfer rate constant for the process occurring at more positive potentials be much less than that for the first process, which causes the second peak to occur increasingly positive of its standard potential as the scan rate increases. As stated above, it was suggested that the small heterogeneous electron transfer rate constants for H_{ae} are connected with the gauche lone pair geometry of these conformations (5b). (It will be recalled that slow electron transfer in the *vic*-dibromides was also associated with gauche geometry.)

Kinetics for Conformational Interconversions
By Cyclic Voltammetry

The relatively good fit of the digital simulations to the experimental voltammograms indicated that Scheme 2, in which the conformational interchange is independent of the presence of the electrode, may be an adequate description. If this were true, the equilibria and kinetics for the conformational change could be extracted from the digital simulation data.

As discussed above, the electron transfer kinetic parameters are of very limited significance because of resistance problems at low temperatures. Nevertheless, information about the kinetics of the conformational change could be reasonably accurate. For this purpose, the significant parameter obtained from the digital simulations is the kinetic parameter, $\kappa = a^{\frac{1}{2}}/K_{eq}k_t^{\frac{1}{2}}$, where $a = vnF/RT$, $K_{eq} = [H_{ee}]/[H_{ae}]$, and $k_t = k(H_{ee} \rightarrow H_{ae}) + k(H_{ae} \rightarrow H_{ee})$. This kinetic parameter yields $K_{eq}\sqrt{k_t}$ directly. Table II contains the data for the most intensively studied case, 6, at several temperatures. The regular change in κ as the temperature is lowered, and reasonable constancy of $K_{eq}\sqrt{k_t}$ at four scan rates at a single temperature (see Figure 2), is gratifying. These electrochemical data do not allow separation of K_{eq} from k_t. The conformational change is so rapid that at all temperatures used, the relative sizes of oxidation peaks 1 and 2 were still changing with scan rate. In such regions, a significant value for κ can be obtained, but not for K_{eq}.

As mentioned above, the photoelectron spectrum of 6 showed that a few percent of H_{ee} conformations are present at room temperature. From the kinetic parameters given in Table II, if $K_{eq} = [H_{ee}]/[H_{ae}] = 0.11$ (10% H_{ee} at equilibrium) $\Delta G^{\ddagger}(H_{ee} \rightarrow H_{ae}) = 11.14$ kcal/mole (average of the data at the five temperatures studied, $-55 \pm 10°$, deviation in calculated values ± 0.03); the corresponding numbers for 5% and 15% H_{ee} at equilibrium are 10.47

and 11.56 kcal/mole, respectively. The $\Delta G^{\ddagger}$ ($H_{ee} \rightarrow H_{ae}$) at $-55°$ for
the dimethylated analogue, 12 is 10.4 kcal/mole (2c) determined by
NMR (in acetone). The presence of the extra methyl in 6 should
not raise the activation energy for the ring reversal which is the
highest barrier between H_{ee} and H_{ae} more than a couple of tenth's
of a kcal/mole, so the agreement of the $\Delta G^{\ddagger}$ calculated from the
electrochemical data with that derived from ^{13}C-NMR is remarkably
good, especially considering the great difference in the types of
experiment performed.

     Simulation gave a $K_{eq}\sqrt{k_t}$ value of 1.3 at $-55°$ for tetrahydro-
pyridazine 7. Anderson (29) showed that the lowest of the higher
energy nitrogen inversion and ring reversal (these which cross the
central line of the conformational diagram in Figure 3) is about
12 kcal/mole, considerably higher than that of the lower energy
nitrogen inversion, which converts 7_{ae} to 7_{ee}. Nelsen and Weisman
redetermined the latter barrier by ^{13}C-NMR, obtaining a $\Delta G^{\ddagger}$ ($-55°$)
of 8.0 kcal/mole (2d). An energy diagram for conversion of 7_{ee} to
7_{ae} is shown in Figure 4. The higher barrier will either be point
A (corresponding to a $\Delta G^{\ddagger}$ ($7_{ee} - 7_{ae}$) of ($8-\Delta G°$ ($7_{ee}-7_{ae}$)) or B, the
activation energy for the $7_{ae} \rightarrow 7_{aa}$ ring reversal. This latter
number cannot be directly measured since neither 7_{ee} nor 7_{aa} is
present in detectable concentrations, but it is probably quite near
the 5.4 kcal/mole observed for cyclohexene (30). The presence of
the nitrogens should not appreciably affect the barrier because
their lone pairs are nearly perpendicular at the transition state.
As an excellent analogy, the lower ring reversal barrier of 12 (2c)
is 10.5 kcal/mole ($-67°$), close to the 10.2 ($-67°$) measured for
cyclohexane (31).

TABLE II

Kinetic Parameters for 6 from Digital Simulation of the CV Curves

Temp (°C)	v(V/sec)	κ	$K_{eq}\sqrt{k_t}$
-45	0.10	1.90	1.16
-50	0.10	2.50	0.89
-55	0.05	2.34	0.68
-55	0.10	3.31	0.68
-55	0.20	4.60	0.68
-55	0.50	7.41	0.68
-60	0.10	5.00	0.46
-65	0.10	6.50	0.35

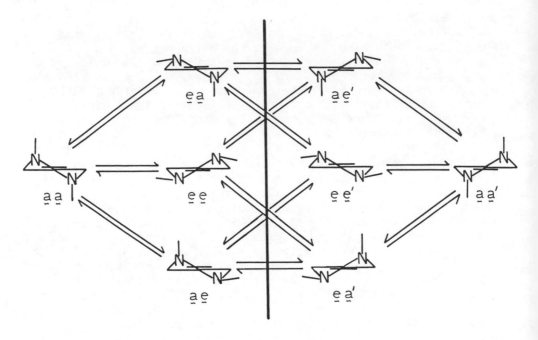

Figure 3.   Conformational Diagram Showing the Interconversions Between the Eight Conformations of 7 (5b).

The fact that $\Delta G°(7_{ee} - 7_{ae})$ must be substantial to preclude detection of 7_{ee} makes it exceedingly likely that point B in Figure 4 is above A, making the lower energy ring reversal the rate limiting step in $7_{ee} \to 7_{ae}$ conversion. Combined with the $K_{eq}\sqrt{k_t}$ value of 1.3, a $\Delta G^{\ddagger}$ $(7_{ee} \to 7_{ae})$ of 5.5 kcal/mole gives a calculated $\Delta G°(7_{ee} - 7_{ae})$ of 3.5 kcal/mole. Even if $\Delta G^{\ddagger}(7_{ee} \to 7_{ae})$ were higher than the 5.5 kcal/mole estimated, $\Delta G°(7_{ee} - 7_{ae})$ is still calculated to be large; an activation energy of 7 kcal/mole gives $\Delta G°$ of 2.7 kcal/mole (corresponding to 0.2% 7_{ee} at -55°).

The value of $K_{eq}\sqrt{k_t}$ obtained is entirely consistent with the failure to detect 7_{ee} and combined with a reasonable estimate for the lower energy ring reversal barrier, gives a probable $\Delta G°(7_{ee} - 7_{ae})$ of over 3 kcal/mole.

The $K_{eq}\sqrt{k_t}$ value obtained by simulation of the -29° cv of 8 was 0.73. By 13C-NMR, interconversion of the mirror image H_{ae} forms, which is likely to proceed through H_{ee} as an intermediate, has $\Delta G^{\ddagger}$ (-29°) of 11.8 kcal/mole (2c). Thus $\Delta G^{\ddagger}(8_{ee} \to 8_{ae})$ might well be 11.8-$\Delta G°(8_{ee} - 8_{ae})$. It is also possible that the lowest energy pathway for the conversion would instead involve the ring

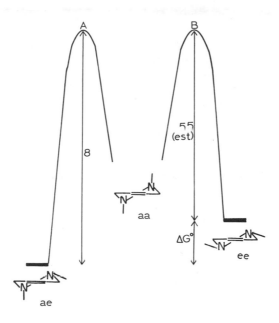

Figure 4.   Schematic Energy Diagram for the Interconversion
of 7_{ae} and 7_{ee} (5b).

flip which gives 8_{aa}.   The activation energy for this process in
the dimethyl compound 12 is 10 kcal/mole, and the methyl substitution is not expected to change the activation energy much ($2c$).
Use of a barrier of 10 kcal/mole in combination with the $K_{eq}\sqrt{k_t}$
value gives a $\Delta G^{\circ}(8_{ee} - 8_{ae})$ of 2.2 kcal/mole (1% ee at -29°); if
a lower energy pathway were available, a higher $\Delta G^{\circ}(8_{ee} - 8_{ae})$
value would be calculated.   Thus the $K_{eq}\sqrt{k_t}$ value observed is
consistent with not detecting 8_{ec}.

Comparison of Conformational Analysis By
Cyclic Voltammetry and NMR

The electrochemical technique is of considerable utility in
obtaining information about conformational equilibria and kinetics
of these hydrazines.   The information obtained is complementary to,
although often different from, that obtainable by the established
technique of low temperature NMR.   The power of low temperature
NMR is well illustrated by consideration of the conformations of
12 (the conformational diagram for 12 (2a,2c) is identical to that
shown in Figure 3, except of course, chair instead of half-chair
conformations are involved).   ^{1}H-NMR experiments allow determination of the rate of the slower of the two processes, nitrogen

inversion which interconverts 12_{ea} and $12_{ee'}$ conformations, or ring reversal which converts 12_{ea} and $12_{ae'}$ (that is, the equilibria crossing the heavy line in the analogue of Figure 3), while ^{13}C-NMR experiments allow determination of both the $12_{ee} \rightleftarrows 12_{ea}$ rates (and K_{eq}) and the $12_{ae} \rightleftarrows 12_{ea}$ rate, in different temperature ranges   The cv experiment, in contrast, is only capable of distinguishing the rapidly oxidizing H_{ee} conformations from the slowly oxidizing H_{ae} conformations, so it can only study the $H_{ee} \rightleftarrows H_{ae}$ equilibrium.

For many six-ring hydrazines (7, 8 and 9 for example), however, H_{ee} conformations are not present in detectable concentration at low temperatures, and then no kinetic information about the $H_{ee} \rightleftarrows H_{ae}$ equilibrium can be obtained from NMR experiments. Such information can, however, still be obtained by cv, as is discussed in detail above for the case of 7. When only H_{ee} conformations are populated, neither cv nor NMR experiments give kinetic data concerning the $H_{ae} \rightleftarrows H_{ee}$ equilibrium.

The simplification of the cv experiment, which only distinguishes conformations on the basis of their electron transfer rates, is welcomed for an unsymmetrical compound like 6 (four 6_{ae} and two 6_{ee} conformations are theoretically possible and at least three are significantiy populated (2d)), for which the low temperature ^{13}C-NMR spectrum was so complex that no conformational information at all was obtained (2d). The fact that the cv experiment is performed on at least 100-fold less material than the ^{13}C-NMR experiment can also be of considerable practical advantage.

In principle, K_{eq} can be measured by cv simply by running the voltammogram at low enough temperature and at fast enough scan rates so that the relative sizes of the two oxidation peaks no longer vary with scan rate. Under these conditions, k_t has been "frozen out" (that is, the $H_{ae} \rightleftarrows H_{ee}$ interconversion is slow on the time scale of the cv experiment), and the simulation gives K_{eq} directly. This was not practical for 6-9 because the effects of solution resistance were serious below -60°, tending to broaden the oxidation peaks and obscure their resolution, and resistance problems also become increasingly great as the scan rate is increased. For 6, even at -60°, the relative sizes of the two oxidation peaks were still not constant at 2 and 5 V/sec, showing that K_{eq} was smaller than the 0.3 obtained by comparing the relative sizes at 5 V/sec, but not allowing a lower limit to be placed on K_{eq}.

In very recent work (18) on hydrazines having $H_{ae} \rightleftarrows H_{ee}$ barriers above 12 kcal/mole (including 13 and five 1,6-diazabicyclo [4.4.0]decane derivatives), "frozen" voltammograms were obtained in the 0° to -50° range, and it was established that the cv method

gives very similar equilibrium constants to those determined by
NMR experiments.

5. CONCLUSIONS

The data for electrochemical oxidation of six-ring tetraalkylhydrazines is adequately reproduced by the model given in Scheme
2, in which H_{ee} undergoes relatively rapid, nearly electrochemically
reversible electron transfer to give the radical cation (H^+), but
H_{ae} gives slow, electrochemically irreversible electron transfer
to give the same cation. An alternate scheme, in which H_{ae} is
first oxidized to an unstable cation of geometry similar to H_{ae},
which rapidly relaxes to H^+, is kinetically equivalent and cannot
at present be distinguished. The most valuable parameter derived
from the model in Scheme 2, $K_{eq}\sqrt{k_t}$, was found to be consistent with
spectroscopic studies of the hydrazines (5b).

It is interesting to extrapolate the conclusions of Scheme 2
to room temperature. For compound $\underline{6}$, the rates of conformational
change increase as the temperature increases, and at constant scan
rate, the first peak grows at the expense of the second. Finally,
a temperature is reached where a single, relatively reversible,
voltammetric peak is obtained. All of the $\underline{6}$ which diffuses to the
electrode is rapidly oxidized, but the stabler H_{ae} conformations
are converted to H_{ee} conformations before electron transfer occurs.
This is a significant conclusion, because it is the first case
where a diffusion-controlled electrode reaction has been shown to
proceed by way of a particular reactant configuration.

The ability to see two oxidation peaks for six-ring hydrazines
is associated with the relatively large activation energies for
conformational interconversion in these compounds and the substantial differences in the electron transfer rate constants for the
two conformations. Whether other classes of compounds will meet
these requirements remains to be seen.

6. SUMMARY

Possible conformational effects in electrochemistry are discussed with emphasis on the question of what factors are necessary
to detect separate voltammetric peaks for different conformers of
a molecule. Earlier electrochemical studies relevant to this
point are reviewed and the results of cyclic voltammetric studies
of the oxidation of tetraalkylhydrazines are discussed in detail.
Certain six-ring cyclic tetraalkylhydrazines show two oxidation
peaks at low temperature and rapid scan rate. The less positive
peak is due to oxidation of conformations with diequatorial alkyl
groups while the more positive peak is due to those with axial,

equatorial alkyl groups. The data have been analyzed quantitatively giving information about the conformational rates and equilibria and the conclusions are consistent with the results of low temperature NMR studies. The cyclic voltammetric technique for conformational analysis is compared with low temperature NMR with respect to the contrasting types of information which can be obtained.

7. ACKNOWLEDGMENTS

This research was supported by the National Science Foundation (grants CHE75-04930 and MPS74-19688) as well as the Wisconsin Alumni Research Foundation. We are pleased to thank our able coworkers who worked out the conditions for experimentally obtaining useful low temperature cv data on tetraalkylhydrazines, Luis Echegoyen and Edward L. Clennan.

8. REFERENCES

(1a)   S. F. Nelsen and J. M. Buschek, J. Am. Chem. Soc., 95, 2011 (1973).

(1b)   S. F. Nelsen, J. M. Buschek and P. J. Hintz, ibid., 95, 2013 (1973).

(1c)   S. F. Nelsen and J. M. Buschek, ibid., 96, 2392 (1974).

(1d)   S. F. Nelsen and J. M. Buschek, ibid., 96, 6982 (1974).

(1e)   S. F. Nelsen and J. M. Buschek, ibid., 96, 6987 (1974).

(2a)   S. F. Nelsen and G. R. Weisman, J. Am Chem. Soc., 96, 7111 (1974).

(2b)   S. F. Nelsen and G. R. Weisman, ibid., 98, 1842 (1976).

(2c)   S. F. Nelsen and G. R. Weisman, ibid., 98, 3281 (1976).

(2d)   G. R. Weisman and S. F. Nelsen, ibid., 98, 7007 (1976).

(3a)   S. F. Nelsen, J. Am. Chem. Soc., 88, 5666 (1966).

(3b)   S. F. Nelsen and P. J. Hintz, ibid., 93, 7105 (1971).

(3c)   S. F. Nelsen, G. R. Weisman, P. J. Hintz, D. Olp and M. R. Fahey, ibid., 96, 2916 (1974).

(3d)   S. F. Nelsen and L. Echegoyen, ibid., 97, 4930 (1975).

(4a) S. F. Nelsen and P. J. Hintz, J. Am. Chem. Soc., 94, 7108 (1972).

(4b) S. F. Nelsen, V. Peacock and G. R. Weisman, ibid., 98, 5269 (1976).

(5a) S. F. Nelsen, L. Echegoyen and D. H. Evans, J. Am. Chem. Soc., 97, 3530 (1975).

(5b) S. F. Nelsen, L. Echegoyen, E. L. Clennan, D. H. Evans and D. A. Corrigan, ibid., 99, 1130 (1977).

(6)   R. Dietz and M. E. Peover, Disc. Faraday Soc., 45, 154 (1968).

(7)   A. J. Bard, V. J. Puglisi, J. V. Kenkel and A. Lomax, Disc. Faraday Soc., 56, 353 (1973).

(8)   L. S. Yeh and A. J. Bard, J. Electroanal. Chem., 70, 157 (1976).

(9)   R. D. Rieke, H. Kojima and K. Öfele, J. Am. Chem. Soc., 98, 6735 (1976).

(10) S. W. Feldberg and L. Jeftic, J. Phys. Chem., 76, 2439 (1972).

(11) Z. R. Grabowski and M. S. Balasiewicz, Trans. Faraday Soc., 64, 3346 (1968).

(12) Z. R. Grabowski, B. Czochralska, A. Vincenz-Chodkowska and M. S. Balasiewicz, Disc. Faraday Soc., 45, 145 (1968).

(13) M. E. Peover, Disc. Faraday Soc., 45, 177 (1968).

(14) Z. R. Grabowski and M. S. Balasicwicz, Disc. Faraday Soc., 45, 178 (1968).

(15) J. Závada, J. Krupička and J. Sicher, Collect. Czech. Chem. Commun., 28, 1664 (1963).

(16) J. Casanova and H. R. Rogers, J. Org. Chem., 39, 2408 (1974).

(17) R. P. Van Duyne and C. N. Reilley, Anal. Chem., 44, 142 (1972).

(18) E. L. Clennan and S. F. Nelsen, unpublished results, 1976.

(19) M. S. Shuman and I. Shain, Anal. Chem., 41, 1818 (1969).

(20) S. Feldberg in "Electroanalytical Chemistry," Vol. 3, A. J. Bard, ed., Marcel Dekker, New York, 1969, pp. 199-295.

(21)   I. Ružić and S. Feldberg, J. Electroanal. Chem., 50, 153
       (1974).

(22)   D. H. Evans, unpublished calculations, University of
       Wisconsin, 1973.

(23)   R. S. Nicholson and I. Shain, Anal. Chem., 36, 706 (1964).

(24)   R. S. Nicholson and I. Shain, unpublished calculations,
       University of Wisconsin, 1963.

(25)   D. H. Evans, J. Phys. Chem., 76, 1160 (1972).

(26)   R. S. Nicholson, Anal. Chem., 37, 667 (1965).

(27)   R. S. Nicholson, Anal. Chem., 37, 1351 (1965).

(28)   H. W. VandenBorn and D. H. Evans, Anal. Chem., 46, 643 (1974).

(29)   J. E. Anderson, J. Am. Chem. Soc., 91, 6374 (1969).

(30)   F. A. L. Anet and M. Z. Haq, J. Am. Chem. Soc., 87, 3147
       (1965).

(31)   F. A. L. Anet and A. J. R. Bourn, J. Am. Chem. Soc., 89, 760
       (1967).

ELECTROCHEMICAL SYNTHESIS AND CHARACTERIZATION OF

TETRATHIOETHYLENES IN NONAQUEOUS SOLVENTS

J. Q. Chambers, R. M. Harnden, and N. D. Canfield

Department of Chemistry, The University of Tennessee

Knoxville, Tennessee   37916

1.   INTRODUCTION

The chemistry of tetrathioethylenes (TTEs) is dominated by their electron donor properties and reversible step-wise oxidation to radical cations and dications (Equation 1). These electron-rich

olefins give two-step oxidation-reduction systems, which places them in the class of molecules which has been expanded and thoroughly studied by Hünig and coworkers (1). Accordingly, voltammetric oxidation of a tetrathioethylene in nonaqueous solvents such as acetonitrile, dimethylsulfoxide, nitromethane, methylene chloride, etc., generally produces two oxidation waves identical with the theoretical shapes of reversible electron transfer reactions. Early examples of this behavior for TTEs can be found in the papers of Geske and Merritt (2), Chambers and coworkers (3, 4), and Hünig et al (5). Known half-wave potentials for TTEs and some selenium analogs are given below.

Interest in these molecules has been heightened considerably in recent years by the discovery that the charge transfer salt of tetrathiafulvalene (TTF) and tetracyanoquinodimethane (TCNQ)

TTF

TCNQ

possesses high electrical conductivity and metallic properties over
a wide temperature range (6,7). When acetonitrile solutions of
TTF, a good electron donor ($E_{1/2}$ = 0.33 V vs. S.C.E.[4]), and TCNQ,
a good electron acceptor ($E_{1/2}$ = 0.20 V vs. S.C.E.[8]), are allowed
to diffuse together, dark-green crystalline needles are formed.
Crystal structure studies (9) show that this salt has a chain
structure with separate stacks of TTF and TCNQ radical ions and
molecules. The physical properties of this salt have been studied
extensively and it is generally accepted that there is only partial
electron transfer between the stacks, each of which contains 50 to
75% radical ions on the average, and both of which contribute to
the electrical conductivity of the salt along the chain axis. Room
temperature conductivities as large as 660 $\Omega^{-1}cm^{-1}$ have been
observed (10,11), and as the temperature is lowered, the conductivity increases in a metallic fashion until 53°K where the TCNQ
stack becomes resistive first, followed by the TTF stack at 38°K
(12,13).

Substitution of selenium for sulfur in the TTF structure to
give tetraselenafulvalene, TSeF (14), results in a donor molecule
which has a greater polarizability and orbital extent than TTF.
This modification of the donor results in an increase in the
metallic conductivity of the resulting charge-transfer salt in
some cases. For example, the room temperature conductivity for
hexamethylenetetraselenafulvalene-TCNQ is approximately 2×10^3
$\Omega^{-1}cm^{-1}$, the largest conductivity reported to date for an "organic
metal" (15).

In addition to the TCNQ salts, recent studies on the solid
state conductivity and crystal structure of TTF halide salts have
revealed the existence of new materials containing metallic
segregated stack phases of the type found in conducting TTF-TCNQ
crystals (16-23). These structures, which often have variable
compositions and are difficult to purify by recrystallization,
have been identified for the mixed-valence nonstoichiometric
TTF-halide salts, TTFX$_a$ (X = Cl, Br, I and 0 < a < 1). In contrast
to these salts, the known stoichiometric TTFX or TTFX$_2$ salts are
insulators. For example, the TTFBr$_{0.7-0.8}$ salt is 13 orders of
magnitude more conducting than the stoichiometric TTFBr salt (18,
24).

Typically these halide salts have been prepared by halogen
oxidation of the neutral donor or by metathetical reaction of TTF
salts in appropriate solvents (25). These methods do not permit
the control over the stoichiometry of the TTFX$_a$ salts which can be
realized by electrosynthesis via generation of the donor radical

cations followed by metathetical reaction with halide salts (26). The salts can also be prepared by photooxidation of the donors in CX_4 solutions (23).

The radical cations and dications of tetrathioethylenes are far more susceptible to chemical reactions such as nucleophilic attack and fragmentation than the neutral parents. This is evident in the voltammetric behavior, which is presented below, of TTF, TSeF, and some TTF derivatives in the presence of bromide ion. These experiments along with potentiometric measurements in tetra-alkylammonium tribromide-TTF acetonitrile solutions provide further insight into the formation and properties of these highly interest-ing radical cation salts.

2.   RESULTS AND DISCUSSION

Half-wave Potentials.   Known half-wave potentials for tetra-thioethylenes are collected in Table 1. Where possible the poten-tials are adjusted to a "S.C.E. scale" in order to facilitate comparisons. A wide range of electron donating ability is indicated for the TTE's in Table 1; the half-wave potentials for the first electron transfer step vary from 1.18 V for tetraphenyltetrathio-ethylene to 0.24 vs. S.C.E. for tetramethyltetrathiafulvalene. This corresponds to an ionization potential range of 6.9 to 7.8 eV using Parker's relation, IP = $E_{1/2}$ + 6.70 (27).

The extent of the pi electron network available for charge delocalization markedly influences the half-wave potentials. This effect is most clearly seen for TTF and its dihydro and tetrahydro derivatives (compounds 10, 9 and 8 in Table 1) (4). The potentials increase by 0.09 V and 0.39 V when the 1,3-dithiole rings are hydrogenated.

Substitution of selenium for sulfur in the TTF structure to give tetraselenafulvalene (TSeF) decreases the electron donating ability by 0.1 eV. This decrease has been attributed to weaker carbon-selenium pi bonds for -C = $\overset{+}{Se}$- in $TSeF^{\ddagger}$ than carbon-sulfur pi bonds for -C = $\overset{+}{S}$- in $TTF^{\ddagger}$ (28). Accordingly, the positive charge is more extensively delocalized in $TTF^{\ddagger}$ than in $TSeF^{\ddagger}$ and $E_{1/2}^{1}$ is lower.

The difference between the half-wave potentials, $\Delta E_{1/2} = E_{1/2}^{2}$ - $E_{1/2}^{1}$, is closely related to coulombic interactions in the dications. Minimization of interactions between positive sulfur atoms decreases $\Delta E_{1/2}$ and consequently increases the dispropor-tionation equilibrium constant (K_{dispro}) of the radical cation. This trend is evident in the data of Table 1 in several instances The smallest $\Delta E_{1/2}$ values are observed for the tetrathioalkyl

Table 1

Voltammetric Half-Wave Potentials of Tetrathio- and Tetraselenoethylenes

$$\begin{array}{c} R_4X \\ R_3X \end{array} {=} \begin{array}{c} XR_1 \\ XR_2 \end{array} \qquad X = S, \text{ unless noted otherwise}$$

No. Compound	$E^1_{1/2}$ (V vs. s.c.e.)	$E^2_{1/2}$ (V vs. s.c.e.)	Solvent/Electrolyte	Ref
1 R_1-R_4 = -CH_3	0.89[a]	0.89[a]	0.1M TEAP/MeCN	2
2 R_1-R_4 = -C_2H_5	0.93[a]	0.93[a]	0.1M TEAP/MeCN	2
3 R_1-R_4 = -$CH(CH_3)_2$	0.97[a]	0.97[a]	0.1M TEAP/MeCN	2
4 R_1-R_4 = -C_6H_5	1.18	1.4[b]	0.1M TEAP/MeCN	2,4
5 $R_{1,4}$ = -$CH_2CH_2CH_2$- $R_{2,3,}$ = -C_2H_5	0.86[a]	0.86[a]	0.1M TEAP/MeCN	47
6 $R_{1,4}$, $R_{2,3}$ = -$CH_2CH_2CH_2$-	0.90	1.04	0.1M TEAP/MeCN	47
7 $R_{1,2}$, $R_{3,4}$ = -$CH_2CH_2CH_2$-	0.73	0.90	0.1M TEAP/MeCN	47
8 $R_{1,2}$, $R_{3,4}$ = -CH_2CH_2-	0.71	1.17	0.1M TEAP/MeCN	47
9 $R_{1,2}$ = -CH_2CH_2- $R_{3,4}$ = -CH=CH-	0.41	0.89	0.1M TEAP/MeCN	4
10 $R_{1,2}$, $R_{3,4}$ = -CH=CH-	0.33	0.70	0.1M TEAP/MeCN	4
11 $R_{1,4}$, $R_{2,3}$ = -CH_2CH_2-	0.90	1.28	0.1M TEAP/MeCN	47
12 $R_{1,4}$, $R_{2,3}$ = -CH=CH-	0.90[c]	1.30[c]	0.05M TEAP/MeCN	46
13 $R_{1,2}$, $R_{4,3}$ = -C=CH- (C$_6$H$_5$)	0.382	0.745	0.05M TEAP/MeCN	49
14 $R_{1,2}$, $R_{3,4}$ = -C=CH- (p-$CH_3OC_6H_4$)	0.340	0.705	0.05M TEAP/MeCN	49
15 $R_{1,2}$, $R_{3,4}$ = -C = C- (CH_3 CH_3)	0.238	0.610	0.05M TEAP/MeCN	49

Table 1 (Continued)

No.	Compound	$E^1_{1/2}$ (V vs. s.c.e.)	$E^2_{1/2}$ (V vs. s.c.e.)	Solvent/Electrolyte	Ref
16	$R_{1,2}, R_{3,4} =$ $H_3CH_2CH_2C\ CH_2CH_2CH_3$, $-C=C-$	0.23^c	0.61^c	0.1M TEAP/MeCN	50
17	$R_{1,2}, R_{4,3} =$ $C_6H_5\ CH_3$, $-C=C-$	0.315	0.690	0.05M TEAP/MeCN	49
18	$R_{1,2}, R_{3,4} =$ $C_6H_5\ C_6H_5$, $-CH\!\!-\!\!CH-$	0.79	1.00	0.1M TEAP/MeCN	51
19	$R_{1,2}, R_{3,4} =$ $CN\ CN$, $-C=C-$	1.12	1.22	0.05M TEAP/MeCN	49
20	$R_{1,2}, R_{3,4} =$ $F_3C\ CF_3$, $-C=C-$	1.05	1.28	0.1M TBuAP/MeCN	52
21	$R_{1,2}, R_{3,4} =$ $CH_3S\ SCH_3$, $-C=C-$	0.47	0.71	0.05M TEAP/MeCN	45
22	$R_{1,2}, R_{3,4} =$ (cyclohexene ring)	0.69^c	1.09^c	0.1M TEABF$_v$/MeCN	5
23	$R_{1,2}, R_{3,4} =$ (benzene ring)	0.61	0.93	0.1M TEAP/MeCN	57
24	$R_{1,2}, R_{3,4} =$ (methyl-substituted ring), CH_3	0.97^c	1.31^c	0.1M TEABF$_v$/MeCN	5
X = Se					
25	$R_{1,2}, R_{3,4} = -HC=CH-$	0.48	0.76	0.1M TEAP/MeCN	28

Table 1 (Continued)

No.	Compound	$E^1_{1/2}$ (V vs. s.c.e.)	$E^2_{1/2}$ (V vs. s.c.e.)	Solvent/Electrolyte	Ref
26	$R_{1,2}$, $R_{3,4}$ = H_3C CH_3 $-C = C-$	0.54	0.93	0.2M TBuABF$_4$/CH$_2$Cl$_2$	53
27	$R_{1,2}$, $R_{3,4}$ =	0.55	0.92	0.2M TBuABF$_4$/CH$_2$Cl$_2$	53
28	$R_{1,2}$, $R_{3,4}$ = H_3CSe $SeCH_3$ $-C = C-$	0.58	0.81	0.1M TEAP/MeCN	54
29	X = S, Se and	0.40	0.72	0.1M TEAP/MeCN	28
30		0.28[d]	0.63	TEAP/MeCN	55

[a] estimated: waves overlap
[b] estimated: second wave is irreversible
[c] Adjusted to S.C.E. scale using Table I of reference 56.
[d] quasi-reversible

derivatives in which the sulfur atoms are free to move so as to decrease the coulombic interactions.

For TTEs with groups bridging the sulfur atoms, the positively charged ring sulfur atoms are forced to interact in the dication. For ethylene or vinylene bridges, $\Delta E_{1/2}$ is ca. 0.4 V; while for trimethylene bridging groups, which permit a larger S-C-S angle, values of 0.17 V and 0.14 V are found. Substitution of electron donating groups on TTF also decreases $\Delta E_{1/2}$. For example, $\Delta E_{1/2}$ is 0.27 V for tetrathiomethoxy-TTF which is 0.07 V less than $\Delta E_{1/2}$ for TTF. Interestingly, it has been proposed that large K_{dispro} values (small $\Delta E_{1/2}$) favor the electron hop process in the conducting salts (29).

Reactions with water. The general features of peak voltammograms of TTE molecules are usually dependent on the presence of trace amounts of nucleophiles such as water in the electrochemical solvents. For example, addition of water decreases the apparent diffusion coefficient of 8 in acetonitrile solutions, shifts E_p^2 to more negative potentials, and increases the irreversibility of the second wave. The latter effect is manifested by the peak current ratio, i_p^c/i_p^a, for the second wave which becomes less than unity in the presence of trace amounts of water. These phenomena are evident in the cyclic voltammograms of 8 shown in Figure 1 (30). The half-wave potential and the reversibility of the first wave are not changed by water concentrations up to 1 M.

These results are consistent with an EEC mechanism in which the dication reacts with water to give follow-up products (Equation 2).

$$8 \underset{\longleftarrow}{\overset{-e^-}{\longrightarrow}} 8^{+} \underset{\longleftarrow}{\overset{-e^-}{\longrightarrow}} 8^{2+} \overset{H_2O}{\underset{k_f}{\longrightarrow}} \text{products} \qquad (2)$$

For this mechanism, the peak potential of the second wave should be dependent on sweep rate and attain the reversible value at the faster sweep rates. The rate constant, k_f, can be evaluated from the shift of the peak potential from the reversible value, E_p^r, using Equation 3 (31).

$$E_p - E_p^{\ r} = \frac{RT}{F} \{0.780 - 0.5 \ln(k_f/a)\} - 0.0285 \qquad (3)$$

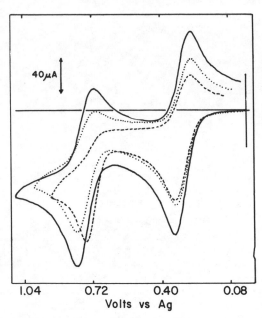

Figure 1.   Cyclic voltammograms of 8 in 0.1M TEAP, CH_3CN; solid line: 1.49x10^{-3}M 8, "anhydrous" $C\tilde{H}_3CN$; dotted line: 1.11x10^{-3}M 8, "usual" $\overline{C}H\tilde{}_3CN$; dashed line: 1.10x10^{-3}M 8, 1% water (V:V).

In this equation a is (F/RT)v where v is the sweep rate, an n-value of one is assumed, and the potential shift is given in volts. Sixteen cyclic voltammograms of the system of Figure 1 were ob-tained at various water concentrations in order to evaluate E_p^r in "anhydrous" acetonitrile.   Since the first wave is fully reversible under aqueous or anhydrous conditions, it can be used as an in-ternal potential reference in order to minimize liquid junction potential changes as the water concentration is varied (32,33). In the anhydrous solvent the second wave peak potential, E_p^r, was 0.4523 V vs. the first wave.   Pseudo first-order rate constants measured in this manner for a 0.642 mM solution of 8 are shown in Figure 2 as a function of added water concentration.   The excellent linearity of the plot implies a second order reaction of water and the dication, and a second-order rate constant of 2.7 $\pm$ 0.2x10^2 ℓ mole^{-1} sec^{-1} is calculated from the slope of the line.   A residual water (or any nucleophile) concentration of 0.02 M can be estimated from the non-zero intercept of the straight-line plot.   Similar results have been obtained by other workers for the reaction of water with electrogenerated cations (34).

The suspected dication hydrolysis products are given in Equa-tion 4, but these have not been confirmed by product isolation.

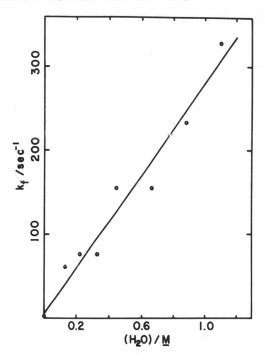

Figure 2. Plot of pseudo first-order rate constant for Equation 1 vs. [H₂O].

Reactions with Halides. The addition of halide salts alters the electrochemical behavior of TTEs more drastically than the addition of water. The cyclic voltammograms of TTF, TSeF, and benzotetrathiofulvalene (BzTTF) in the presence of added tetrabutylammonium bromide demonstrate this effect. The half-wave potentials $(E_{1/2})$ of these donor molecules, TTF (0.33V), TSeF (0.48V), BzTTF (0.61V), in combination with the formal potential for the Br_3^-/Br^- couple in acetonitrile (0.45V), establish that the reaction of the radical cation with bromide (Equation 5) is most favored for BzTTF.

$$2 \ R^{\overset{+}{\cdot}} \ + \ 3Br^- \ = \ 2R \ + \ Br_3^- \tag{5}$$

The voltammetric behavior is in accord with this view in that

cyclic voltammograms of both TSeF and BzTTF exhibit waves which
are related to bromide-radical cation salts, while voltammograms
of TTF exhibit only waves due to TTF and the Br_3^-/Br^-, Br_2/Br_3^-
couples in acetonitrile (35). These reactions may play a role in
the formation of conducting salts since Equation 5 represents a
means by which the donor radical cation chains can be partially
reduced.

For TSeF, addition of TBABr results in the sharpening of both
the first and second TSeF oxidation waves in the cyclic voltammograms and a shift of the waves to more negative potentials.   For
the first TSeF wave at a 1:1 molar ratio of Br^- to TSeF the peak
width decreases from 60 mV to ca. 40 mV and the peak current
function, $i_p/\sqrt{v}$, increases from 160 to 190 μA $V^{-1/2}$ $sec^{1/2}$.   Increased amounts of TBABr shift the first oxidation wave to more
negative potentials, $(\partial E_p/\partial \log C) = -50$ mV, but do not significantly change the shape of the peak voltammogram.

These results suggest the participation of bromide ions in
the electrode reaction and are consistent with formation of a
solid bromide salt on the electrode surface, Equation 6. The solid

$$TSeF \quad + \quad 0.8 \ Br^- \quad \longrightarrow \quad TSeFBr_{0.8}(s) \quad + \quad 0.8e^- \qquad (6)$$

$TSeFBr_{0.8}$, which precipitates when TBABr is added to solutions of
coulometrically generated $TSeF^+$, may be absorbed on the platinum
surface.

The one-electron BzTTF oxidation wave and the Br^-/Br_3^-
oxidation wave are nearly coincident at the 10^{-3} $\underline{M}$ level. Cyclic
voltammograms of BzTTF/TBABr mixtures in acetonitrile exhibit a
reduction "postwave" which is not characteristic of either $BzTTF^+$
or Br_3^- reduction waves. It is likely that this wave is due to
reduction of $BzTTFBr_x$. On repeated cycling the platinum electrode
passivates for reduction of BzTTF oxidation products.

Coulometric oxidation of a 1.6×10^{-3} $\underline{M}$ BzTTF, 0.1 $\underline{M}$ TEAP,
CH_3CN solution consumed 1 F/mole of electricity and resulted in
precipitation of a black perchlorate salt. After dilution to
dissolve the solid, $BzTTFCl_x$ and $BzTTFBr_x$ salts were precipitated
by addition of the respective tetraalkylammonium halides. The
chloride salt was obtained as lustrous black crystals and the
bromide was an amorphouse black solid; the values of x were not
determined. Solubilities of the salts were determined by spectrophotometry using either the 440 nm band ($\log \varepsilon = 3.40$) or the 625
nm band ($\log \varepsilon = 3.26$) of $BzTTF^+$. The bromide salt was about five
times more insoluble than the chloride salt; the latter had a
solubility of 5×10^{-4} moles/liter in acetonitrile at room temperature.   Rough conductivity measurements on the solids with a

volt-ohm meter indicated that $BzTTFCl_x$ was nonconducting ($\sigma < 10^{-6}$ Ω^{-1} cm^{-1}) and that $BzTTFBr_x$ was conducting ($\sigma < 10^{-2}$ $\Omega^{-1}cm^{-1}$). This result is in accord with the view that a positive or not too negative $E°$ for Equation 5 favors formation of the conducting mixed valence compound (26).

Reaction of TTF with tribromide. Although cyclic voltammograms of TTF-TBABr mixtures gave no indication of $TTFBr_x$, potentiometric studies using a platinized platinum electrode permitted the formation of this salt to be followed. Acetonitrile solutions of tribromide which were prepared either by constant potential coulometric oxidation of TBABr or from reagent grade tetramethylammonium tribromide were used as the oxidant. In acetonitrile the tribromide is stable with respect to disproportionation and the $E°$ of the Br_3^-/Br^- couple is 0.147 V vs. $Ag/AgNO_3(0.01\underline{M})$, TEAP ($0.1\underline{M}$). This formal potential, which was determined from cyclic voltammograms of TBABr at several concentrations, indicates that Br_3^- will readily oxidize TTF in acetonitrile.

When the tribromide solutions were added to TTF solutions, a dark red solid formed. The indicated reaction is given in Equation 7 and analysis of the titration curves gave x = 0.79 suggesting

$$2 \; TTF + x \; Br_3^- \longrightarrow 2 \; TTFBr_x + x \; Br^- \qquad (7)$$

formation of the electrically conducting nonstoichiometric $TTFBr_x$ salt. When the solid was collected from solutions containing excess Br_3^- and subjected to X-ray diffraction analysis, both $TTFBr_x$ and TTFBr were found to be present.

The thermodynamic solubility product of $TTFBr_x$ could be

$$K_{sp} = a_{TTF}^{1-x} \; a_{TTF^+}^{x} \; a_{Br^-}^{x} \qquad (8)$$

estimated from the platinum potential measurements in the solutions which contained excess TTF. A value of 4.3×10^{-8} was obtained for K_{sp} using the Debye-Huckel limiting law to estimate activity coefficients for TTF^+ and Br^- and assuming a unity activity coefficient for the neutral TTF. This K_{sp} value is highly dependent on the $E°'$ value used for the TTF^+/TTF couple. A value of 0.03 V vs. $Ag/AgNO_3(0.01\underline{M})$, TEAP($0.1\underline{M}$) was used in the calculation. The activity coefficient correction was small since the ionic strength was less than 10^{-3} $\underline{M}$ for most of the measured potentials.

The platinum electrode potential in solutions containing excess tribromide in contact with solid $TTFBr_x$ was not stable, but decreased over a period of 15 minutes or longer. The reason for

this is believed to be due to a slow heterogeneous formation of the
black insulating TTFBr salt, Equation 9. This is supported by

$$TTFBr_x(s) \xrightarrow{\ Br_3^-\ } TTFBr\ (s) \qquad\qquad (9)$$

X-ray diffraction and chemical analyses of the collected solid.
In a practical vein, this suggests that the appropriate solution
for recrystallization or formation of the $TTFBr_x$ salt contains the
stoichiometric ratios given in Equation 7.

     TTE Dication Rearrangement. Cyclic voltammograms of the en-
docyclic TTEs 6 and 11 reveal an interesting endocyclic to exocyclic rearrangement reaction of the dications 6^{2+} and 11^{2+}.
(This work has appeared in communication form (36).) Cyclic

6   11

voltammograms of both 6 and 11 contain reversible product waves
which grow in magnitude upon repeated cycles in addition to the
usual one-electron TTE couples. These new waves appear only when
the platinum electrode potential is swept into the region where
the dications are generated. Figures 3 and 4 show the behavior
for 6.   The couples Ox1/R1 and Ox2/R2 are the usual one-electron
waves and Ox3/R3 is the new couple. The new couple can be
assigned to the exocyclic TTE couple, $7^{\ddagger}/7$ on the basis of its
wave shape and $E_{1/2}$ value. Comparison of the voltammograms of 6
and 7 reveals that the reversible wave of the exocyclic dication
overlaps exactly with the wave of the endocyclic cation and therefore is not evident in the cyclic voltammograms of 6 in Figure 3.
The voltammograms of 11 show all four waves, however, since the
half-wave potentials of 8 and 11 do not coincide. The suggested
rearrangement is given in Equation 10.

(10)

     Further evidence to support the indicated rearrangement is
presented in the e.s.r. spectrum obtained from solutions of 6. In
situ electrolysis of 6 at +0.92 V generated a green-colored radical
species after ca. 15 minutes. A single nine-line spectrum
(Figure 5) was obtained corresponding to eight equivalent hydrogen
atoms with a coupling constant a_H = 3.19 Gauss and a g-value =
2.0085 in acetonitrile solution.

Although 6^{+} and 7^{+} would be expected to exhibit similar e.s.r. spectra, it seems likely that the e.s.r. spectrum of 7^{+} was obtained when 6 was electrolyzed in the cavity. The radical cation of 7 could be produced as the predominant species <u>via</u> a disproportionation mechanism, Equations 11-13, during the time scale of the <u>in situ</u> electrolysis. In support of this scheme, <u>in situ</u>

$$2\ R^{+}_{endo}\ \rightleftharpoons\ R_{endo}\ +\ R^{2+}_{endo} \tag{11}$$

$$R^{2+}_{endo}\ \longrightarrow\ R^{2+}_{exo} \tag{12}$$

$$R^{2+}_{exo}\ +\ R_{endo}\ \rightleftharpoons\ R^{+}_{exo}\ +\ R^{+}_{endo} \tag{13}$$

electrolysis of 11 in the e.s.r. cavity resulted in a more complex signal, which was initially composed of two nine-line patterns. The disproportionation equilibrium constant for 11^{+} is much smaller than for 6^{+}, 3.7×10^{-7} <u>vs.</u> 3×10^{-3}, and consequently 11^{+} would have

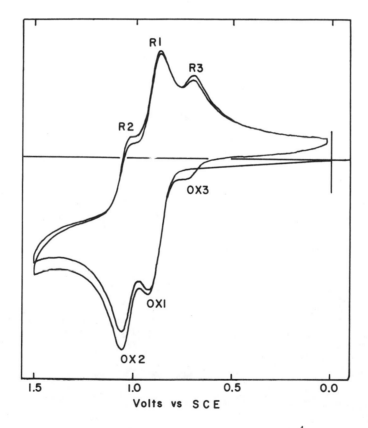

Figure 3. Cyclic voltammogram of 6, 5.0×10^{-4}M, 0.1M TEAP, CH$_3$CN, 0.104 V sec^{-1}.

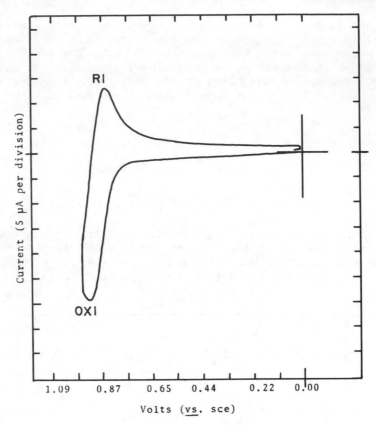

Figure 4. Cyclic voltammogram of 6, 5.0×10^{-4}M, 0.1M TEAP, CH_3CN, 0.104 V sec^{-1}.

a longer lifetime than 6^{+}. After several minutes (ca. 15) the signal due presumably to 11^{+} was gone, and a single nine-line pattern (a_H = 2.26 Gauss, g = 2.0065) remained which can be assigned to the radical cation of 8.

The mechanism of the dication rearrangement is open to speculation, but may involve simultaneous intramolecular migration of two carbon-sulfur sigma bonds of the type described by Reetz (37, 38). Other conceivable mechanisms involve episulfonium ion intermediates of the following type.

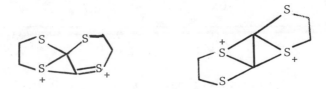

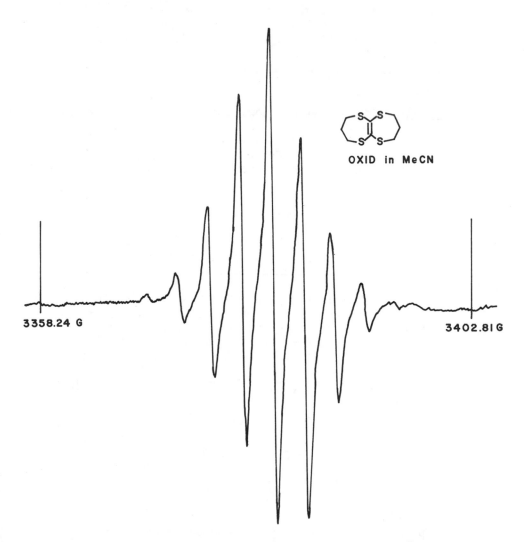

OXID in MeCN

3358.24 G

3402.81 G

Figure 5.   ESR spectrum obtained from 8, $2.0 \times 10^{-3}M$, 0.05M TEAP, CH_3CN, 0.92 V vs. S.C.E.

In these structures the charge is localized on only two of the
four sulfur atoms, which would increase the energy of the transition state relative to that of the dyotropic reaction. The major
driving force for the rearrangement undoubtedly is the minimization of the coulombic, repulsive interactions between positive
sulfur atoms, which is possible by rotation about the central bond
in the exocyclic, but not in the endocyclic dications (36).

3. EXPERIMENTAL SECTION

     Chemicals. For the TTFBr$_x$ work spectrograde acetonitrile
(Burdick and Jackson) was distilled under nitrogen from P_2O_5 prior
to use and was stored in a dry argon atmosphere. For the
CH_3CN/H_2O study a hybridized variation of the O'Donnel et al (39)
and the Walter and Ramaley (40) methods, which was developed in
our laboratory (30,41), was employed.

     The best "anhydrous" solvent was obtained after a final slow
distillation, transfer into the glove box, and immediate use. For
the dication rearrangement work, either the latter method was used
to purify the solvent or spectrograde, "gold label" acetonitrile
(Aldrich Chemical Co.) was used directly.

     The tetrathiafulvalene derivatives, TTF (42), BzTTF (43), and
TSeF (14), were prepared according to previously published methods.
The TTEs 6 and 11 were prepared via the reduction of CS_2 (44) using
a platinum cathode (45) according to the following scheme:

(yield: 40-50%)

6

A similar procedure has recently been reported by Mizuno et al.
(46) who employed alkaline hydrolysis to eliminate CS from the
thione. The chemical method appears to be superior to the electrochemical reduction which gave poor yields at greater than the
millimolar level. Compound 6 2,6,8,12-tetrathiabicyclo[5.5.0]
dodec-1(7)-ene, was obtained as a white crystalline solid after
recrystallization from ethanol; mp 121-122°; nmr (CDC1$_3$) δ 3.1
[triplet, 8H, SCH$_2$], δ 2.2[multiplet, 4H, CH$_2$]. Compound 11,
2,5,7,10-tetrathiabicyclo[4.4.0]dec-1(6)-ene, was obtained as pale
tan-colored crystals after recrystallization twice from absolute
ethanol; mp 152.5 - 153.5°; nmr (CDCℓ$_3$) δ 3.25[s, 8H, CH$_2$]. Mass

spectra and elemental analyses were consistent with the given
structures (47).

   Electrodes. The working electrodes for the voltammetric experiments were platinum disk electrodes which were polished to a
mirror finish with either Buehler 0.5μm or 0.03μm lapping compound.
A large platinum screen electrode was used for the coulometric
generation of the donor radical cations in the halide study and a
platinum screen was also used for the in situ electrolyses in the
e.s.r. cavity (48). The reference electrode was usually a silver
wire immersed in 0.01M AgNO₃, 0.1M TEAP, CH₃CN; an aqueous s.c.e.
was used in some cases. The indicating electrode for the potentiometric measurements was a 1 cm^2 platinum foil which had been
lightly platinized in a 3% chloroplatinic acid solution containing
a trace of Pb(II). It was found that this electrode gave more
stable potential readings than a polished platinum electrode in
acetonitrile.

   Instrumentation. Conventional three-electrode potentiostats
were used for the voltammetric and coulometric studies. The
potentiometric measurements were made with a Data Precision Model
1450 digital voltmeter. Experiments which employed "nonaqueous"
acetonitrile were conducted in a dry, oxygen-free, argon filled
dry box (Vacuum Atmosphere Co., Model HE243 Dry Lab equipped with
a Model 40-1 Dri-Train gas purifying unit). Powder X-ray diffrac-
tiondata were used to identify the donor dication, radical cation,
and halide deficient salts which were isolated from electrolysis
experiments.

4.   SUMMARY

   The electrochemical behavior of tetrathioethylenes, including
the highly interesting tetrathiafulvalene (TTF) derivatives, is
surveyed in this paper. Literature half-wave potentials for 30
tetrathioethylene derivatives are tabulated and discussed. Trends
in the potentials are interpreted in terms of delocalization of
positive charge in the cations and coulombic interactions of the
positive sulfur atoms in the dications. Voltammetric evidence is
presented for nucleophilic reactions of water and halide ions with
tetrathioethylene dications and radical cations respectively. The
electrochemical oxidation of a representative tetrathioethylene in
acetonitrile/water mixtures is shown to follow an EEC mechanism
which permits estimation of both the second order rate constant
for the reaction of the dication with water and the water content
of our "anhydrous" acetonitrile. The thermodynamic solubility
product of the electronically conducting TTFBr$_{0.8}$ salt in acetonitrile is estimated from potentiometric data. Finally electron
spin resonance and electrochemical evidence for a novel endocyclic

to exocyclic rearrangement reaction of two tetrathioethylene dications is presented.

5.   ACKNOWLEDGMENT

The portions of this work which were performed at The
University of Tennessee, Knoxville, were supported by the National
Science Foundation, Research Corporation, and The University of
Tennessee. Stimulating discussions with Dennis Green, Frank
Kaufman and Ed Engler of the I.B.M. T.J. Watson Laboratory in
Yorktown Heights, N.Y., contributed greatly to the work described
above. The technical assistance of Dennis Green is also acknow-
ledged.

6.   REFERENCES

(1)   See S. Hünig and H. C. Steinmetzer, J. Liebigs Ann. Chem.,
      1060, 1090 (1976) for recent papers in this series.

(2)   D. H. Geske and M. V. Merritt, J. Amer. Chem. Soc., 91, 6921
      (1969).

(3)   N. D. Canfield, J. Q. Chambers, and D. L. Coffen, J. Elec-
      troanal. Chem., 24, A7 (1970).

(4)   D. L. Coffen, J. Q. Chambers, D. R. Williams, P. E. Garrett,
      and N. D. Canfield, J. Amer. Chem. Soc., 93, 2258 (1971).

(5)   S. Hünig, H. Schlaf, G. Kiesslich, and D. Scheutzow,
      Tetrahedron Lett., 2271 (1969).

(6)   J. Ferraris, D. O. Cowan, V. Walatka, Jr., and J. H.
      Perlstein, J. Amer. Chem. Soc., 95, 948 (1973).

(7)   L. B. Coleman, J. J. Cohen, D. J. Sandman, F. G. Yamagashi,
      A. J. Garito, and A. J. Heeger, Solid State Commun., 12,
      1125 (1973).

(8)   M. R. Suchanski and R. P. Van Duyne, J. Amer. Chem. Soc., 98,
      250 (1976).

(9)   T. J. Kistenmacher, T. E. Phillips, and D. O. Cowan, Acta
      Crystallogr., Sect. B, 30, 763 (1974).

(10) G. A. Thomas, et al., Phys. Rev. B, 13B, 5105 (1976).

(11) M. J. Cohen, L. B. Coleman, A. F. Garito, and A. J. Heeger, ibid., p 5111.

(12) T. D. Schultz and S. Etemad, ibid., p 4928.

(13) P. M. Chaikin, J. F. Kwak, R. L. Green, S. Etemad, and E. M. Engler, Solid State Commun., 19, 1201 (1976).

(14) E. M. Engler and V. V. Patel, J. Amer. Chem. Soc., 96, 7376 (1974).

(15) A. N. Bloch, D. O. Cowan, K. Bechgaard, R. E. Pyle, R. H. Banks, and T. O. Poehler, Phys. Rev. Lett., 34, 1561 (1975).

(16) F. Wudl, D. Wobschall, and E. J. Hufnagel, J. Amer. Chem. Soc., 94, 670 (1972).

(17) B. A. Scott, J. B. Torrance, S. J. LaPlaca, P. Corfield, D. C. Green, and S. Etemad, Bull. Amer. Phys. Soc., 20, 496 (1975).

(18) S. J. LaPlaca, P. W. R. Corfield, R. Thomas, and B. A. Scott, Solid State Commun., 17, 635 (1975).

(19) D. J. Dahm, G. R. Johnson, F. L. May, M. G. Miles, and J. D. Wilson, Cryst. Struct. Commun., 4, 673 (1975).

(20) J. J. Daly and F. Sanz, Acta Cryst., B31, 620 (1975).

(21) R. B. Somoano, A. Gupta, V. Hadek, T. Datta, M. Jones, R. Deck, and A. M. Hermann, J. Chem. Phys., 63, 4970 (1975).

(22) R. J. Warmack, T. A. Callcott, and C. R. Watson, Phys. Rev. B, 12, 3336 (1975).

(23) B. A. Scott, F. B. Kaufman, and E. M. Engler, J. Amer. Chem. Soc., 98, 4342 (1976).

(24) E. M. Engler, Chem. Technology, 6, 274 (1976).

(25) F. Wudl, J. Amer. Chem. Soc., 97, 1962 (1975).

(26) F. B. Kaufman, E. M. Engler, D. C. Green, and J. Q. Chambers, J. Amer. Chem. Soc., 98, 1596 (1976).

(27) V. D. Parker, J. Amer. Chem. Soc., 96, 5656 (1974); M. E. Peover, Electrochim. Acta, 13, 1083 (1968).

(28) E. M. Engler, F. B. Kaufman, D. C. Green, C. E. Klots, and
      R. N. Compton, J. Amer. Chem. Soc., 97, 2921 (1975).

(29) A. F. Garito and A. J. Heeger, Accts. Chem. Res., 7, 232
      (1974).

(30) N. D. Canfield, Ph.D. Thesis, The University of Tennessee,
      Knoxville, Tennessee, 1974.

(31) R. S. Nicholson and I. Shain, Anal. Chem., 36, 706 (1964).

(32) R. Dietz and B. E. Larcombe, J. Chem. Soc., B, 1369 (1970).

(33) R. E. Sioda, J. Phys. Chem., 72, 2322 (1968).

(34) M. F. Marcus and M. D. Hawley, Biochim. Biophys. Acta, 201,
      1 (1970).

(35) A. I. Popov and D. H. Geske, J. Amer. Chem. Soc., 80, 5346
      (1958); I. V. Nelson and R. T. Iwamoto, J. Electroanal. Chem.,
      7, 218 (1964); F. Magno, G. Mazzocchin, and G. Bontempelli,
      J. Electroanal. Chem., 47, 461 (1973).

(36) R. M. Harnden, P. R. Moses and J. Q. Chambers, J.C.S. Chem.
      Commun., 11 (1977).

(37) M. T. Reetz, Angew. Chem. Int. Ed., 13, 402 (1974).

(38) M. T. Reetz, Tetrahedron, 29, 2189 (1973).

(39) J. F. O'Donnel, J. T. Ayres, and C. K. Mann, Anal. Chem., 37,
      1161 (1965).

(40) M. Walter and L. Ramaley, ibid., 45, 165 (1973).

(41) J. A. Lanning and J. Q. Chambers, ibid., 45, 1010 (1973).

(42) L. R. Melby, H. D. Hartzler, and W. A. Sheppard, J. Org.
      Chem., 39, 2456 (1974); F. Wudl and M. L. Kaplan, ibid.,
      p 3608.

(43) S. Hünig, G. Kiesslich, H. Quast, and D. Scheutzow, J.
      Liebigs Ann. Chem., 310 (1973).

(44) S. Wawzonek and S. M. Heilmann, J. Org. Chem., 39, 511 (1974)

(45) P. R. Moses and J. Q. Chambers, J. Amer. Chem. Soc., 94, 945
      (1974).

(46) M. Mizuno, M. P. Cava, and A. F. Garito, J. Org. Chem., 41
      1484 (1976).

(47) R. M. Harnden, M. S. Thesis, The University of Tennessee,
      Knoxville, Tennessee, 1976.

(48) R. N. Adams, "Electrochemistry at Solid Electrodes," Marcel
      Dekker, Inc., New York, NY (1969).

(49) P. Calas, J. M. Fabre, E. Toreilles, and L. Giral, C. R.
      Acad. Sci. Paris, Ser. C, 280, 901 (1975).

(50) L. B. Coleman, F. G. Yamagishi, A. F. Garito, and A. J.
      Heeger, Phys. Lett., 51A, 412 (1975).

(51) P. R. Moses, Ph.D. Thesis, The University of Tennessee,
      Knoxville, Tennessee, 1974.

(52) H. D. Hartzler, J. Amer. Chem. Soc., 95, 4379 (1973).

(53) K. Bechgaard, D. O. Cowan, and A. N. Bloch, Mol. Cryst.,
      Liq. Cryst., 32, 227 (1976).

(54) E. M. Engler, D. C. Green, and J. Q. Chambers, J.C.S. Chem.
      Commun., 148 (1976).

(55) R. C. Wheland and J. L. Gilson, J. Amer. Chem. Soc., 98,
      3916 (1976).

(56) C. K. Mann in "Electroanalytical Chemistry," Vol. 3, A. J.
      Bard, editor, Marcel Dekker, Inc., New York, p. 64 (1969).

(57) J. Q. Chambers and D. C. Green, unpublished result.

THIN LAYER SPECTROELECTROCHEMICAL STUDIES OF COBALT AND COPPER

SCHIFF BASE COMPLEXES IN N,N-DIMETHYLFORMAMIDE

David F. Rohrbach, Edward Deutsch, and William R. Heineman

Department of Chemistry, University of Cincinnati

Cincinnati, Ohio 45221

1. INTRODUCTION

Since its initial development in 1967 (1), the optically
transparent thin layer electrode (OTTLE) has seen increasing application to the spectroelectrochemical study of chemical compounds.
The cell consists of a thin layer of electrolyte, typically 0.05 -
0.2 mm thick, which is confined adjacent to a transparent electrode. The advantageous features of such a cell are (a) rapid
electrolysis (typically 20-120 sec) of the thin solution layer with
diffusion as the only mode of mass transport to the electrode, and
(b) simultaneous recording of spectra by an optical beam passing
through the transparent electrode.

The OTTLE has been used primarily for the acquisition of
spectra of compounds in various redox states, the state(s) being
determined by the potential applied to the transparent electrode,
and for the measurement of n values by thin-layer coulometry.
Examples of such investigations are: the study of acetylacetonate
ligand exchange reactions for iron(II)-iron(III) in acetonitrile
(2); the reduction of ninhydrin in acetonitrile (UV, visible, IR
spectral study) (3); the oxidation of rubrene in benzene-
acetonitrile (4); the reduction of 1-hydroxy-9,10-anthraquinone,
1,8-dihydroxy-9,10-anthraquinone, and 9,10-anthraquinone in nonaqueous solvents (5,6); the reduction of acetylacetone in acetonitrile (7); the reduction of substituted porphyrins in aprotic
(8) and aqueous (9) media; the reduction of the mixed valence
μ-pyrazine-decaamminediruthenium (II,III) complex (10); the reduction of vitamin B_{12} (11) and the cobalamin coenzyme, 5'-deoxy-
adenosylcobalamin (12). More recently, a spectropotentiostatic
technique has been developed for the determination of redox

177

potentials and n̲ values of biological redox components, such as
cytochrome c̲, which do not exchange electrons with an electrode
(13). This technique utilizes a mediator-titrant to electrochemically couple the biocomponent to the electrode potential. Light-
induced absorption changes accompanying P700 photooxidation in
photosystem-I subchloroplasts, and circular dichroism measurements
on soluble spinach ferredoxin have been reported (14). The ability
to measure kinetic rate constants of homogeneous chemical reactions
coupled to a heterogeneous electron transfer has recently been
demonstrated in several laboratories (15-17). The reactive species
can be rapidly generated by thin layer electrolysis and the subsequent chemical reaction monitored optically.

Perhaps the simplest and most widely used OTTLE is the gold
minigrid microscope slide design which is shown in Figure 1. This
cell is easily fabricated from commercially available gold minigrid
which is the optically transparent electrode; Teflon tape spacers,
which define the thickness of the thin solution layer; glass or
quartz microscope slides; and epoxy, which is coated around the
periphery. The thin layer cell is defined by the volume of solution adjacent to the minigrid, typically 25-40 µl. The bottom edge
of the cell is dipped into a small cup containing the solution of
interest as well as the reference and auxiliary electrodes. Solution is drawn into the cell, by suction application at the top,
where it maintains its level above the minigrid by capillary action. The minigrid is available in a variety of mesh sizes and
materials (gold, nickel, silver, copper, and tungsten). The most
useful has been gold in the 100 wpi (82% T) to 2000 wpi (22% T)
range. "Mercury" minigrids, which extend the negative potential
range of the electrode have been formed by electrodepositing thin
films of mercury on nickel (18) and gold (19) minigrids. Since the
optical transparency of the minigrid is caused by holes in the
micromesh, the OTTLE is transparent throughout the UV-visible-IR
range with the appropriate use of quartz or infrared transparent
optical plates in place of the microscope slides. Several variations of this basic cell design which facilitate oxygen removal
and reduce the required volume of solution have been reported
(5,14,20,21). OTTLEs which utilize a thin film of vapor-deposited
platinum as the transparent electrode have also proven useful
(4,10). Optical measurements have been made in thin layer cells
with non-transparent electrodes by specular reflection techniques
(22).

Since the OTTLE provides a convenient way to characterize
electrochemically active compounds with regard to redox potentials,
n̲ values, and spectra of electrogenerated redox states, this technique was selected for the systematic study of metal ion complexes

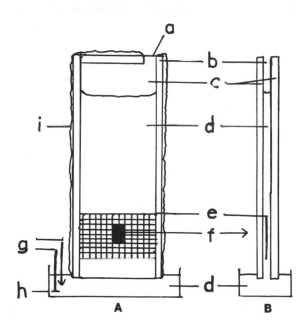

Figure 1.   Optically transparent thin layer electrode.   (A)
Front view.   (B) Side view.   (a) point of application of suction
to change solution, (b) Teflon tape spacers, (c) 1 x 3 in. microscope slides, (d) sample solution, (e) gold minigrid electrode,
(f) optical path of spectrophotometer, (g) reference and auxiliary
electrodes, (h) solution cup, (i) epoxy.

of Schiff bases.   Schiff base complexes are of considerable interest to inorganic chemistry because their structural, chemical,
electrochemical, spectral and electronic properties are often
strongly dependent on the detailed ligand structure, and, therefore, these properties (and their interrelationships) can be systematically examined by varying the nature of the ligand.   Schiff
base complexes of cobalt and copper are also specifically of interest to bioinorganic chemistry since these complexes respectively
provide inorganic models for naturally occurring vitamin B_{12} (23)
and "blue" copper proteins (24).

Using the nomenclature of Holm's extensive 1966 review (25),
the Schiff base complexes of most interest have either the non-
bridged structure

or the bridged structure

where the imine bond $>$C=N- constitutes the essential chemical
feature of this class of compounds. A few systematic electrochemical studies on copper and cobalt Schiff base complexes have been
reported. Calvin and Bailes (26) studied the polarographic reduc-

tion of bridged copper complexes with A = , R = H,

B= -CH$_2$CH$_2$-,   and , as well as some non-

bridged complexes, in aqueous pyridine and found that E$_{1/2}$ was
correlated with the thermodynamic stability of the solution complexes. A polarographic study (27) of non-bridged cobalt

complexes with A = , R = H in aqueous ethylene glycol

revealed that both the metal and ligand are reduced in this protic
environment. In a recent review, Costa (23) has evaluated polarographic E$_{1/2}$ values for a series of cobalt aromatic imine complexes
in terms of electronic and steric effects resulting from the imine
ligand. Most recently, Patterson and Holm (24) have reported an
extensive investigation into the effects of varying Schiff base
structure on the polarographic E$_{1/2}$ values for copper complexes

with A = , CH$_3$-C=CH-, C$_6$H$_5$-C=CH-, and B = -CH$_2$CH$_2$-, ;

other types of copper chelates were also studied. This elegant
study employed N,N-dimethylformamide as solvent and established n
values by controlled potential coulometry. Other workers have
correlated redox potentials for Schiff base complexes with oxygen
binding strength (28) and efficiency of phosphine oxidation

catalysis (29), using the nonaqueous solvents pyridine and acetonitrile.

The overall objective of our work on Schiff base complexes is
to eventually correlate the electrochemical and spectral properties
of a wide variety of these complexes with the electronic and steric
characteristics of variously substituted Schiff base ligands.   To
meet this objective it is first necessary to demonstrate the applicability and efficacy of the OTTLE techniques to this type of
investigation.   This paper describes the application of thin layer
cyclic voltammetry and coulometry, and the application of the
spectropotentiostatic technique (13) for measuring redox potentials and n values, to two typical Schiff base complexes, i.e. the
copper and cobalt complexes of bis(salicylaldehyde)ethylenediimine.

M = Co, Cu

Like most Schiff base complexes, these two are essentially insoluble in water and their imine linkage is potentially susceptible to
hydrolysis.   These properties dictate the use of nonaqueous solvents for studying the chemistry of Schiff base complexes.   After
evaluating both acetonitrile and N,N-dimethylformamide (DMF), we
have chosen to use DMF in this study since (1) it is more readily
purified, (2) it functions as a better Lewis base, a property
which seems to be very important in these systems (30,31), (3) most
Schiff base complexes have significant solubility in DMF, and (4)
DMF has suitable electrochemical and optical properties for the
OTTLE techniques.

2.   EXPERIMENTAL

Apparatus.   The OTTLE was constructed from microscope slides,
2-mil adhesive Teflon tape spacers (Fluorofilm DF-1200 Teflon tape,
Dilectrix Corp., Farmingdale, N.Y.), gold minigrid (500 wires/in.,
60% transmittance, Buckbee Mears Co., St. Paul, Minn.), and epoxy
(1,20).   Strips of Teflon tape ≈2mm wide were pressed along the
periphery of two precleaned (10-15 minutes in a radio frequency
plasma cleaner) microscope slides, as shown in Figure 1.   A 1 x
3.5 cm section of gold minigrid was symmetrically positioned with
tweezers within 3 mm of the bottom edge of one of the slides; such
a location minimized iR drop in the cell.   The second slide

was then laid on top to form a "sandwich" which was clamped to
maintain the proper position during the epoxy application. Epoxy
was then applied along the taped edges, with the pieces of minigrid that extend beyond the cell edges being folded over and held
in place with a dab of epoxy. The epoxy was then cured overnight
at 60-80°C. A piece of metal foil folded over the minigrid tabs
served to protect the grid from tearing and provided electrical
contact between the grid and the alligator clip connected to the
potentiostat. The OTTLE was then supported by an easily fabricated
lucite frame and dipped into the solution cup. This apparatus was
placed in a spectrophotometer cell compartment and positioned so
that the light beam passed directly through the minigrid.

     Cyclic voltammetry and controlled potential coulometry were
performed using a potentiostat of conventional operational amplifier design, an X-Y recorder with time base (Houston Omnigraphic
Model 2000), and a digital voltmeter (United Systems Corp. Model
261C or Fluke Model 8000A). For the spectropotentiostatic exper-
ment a Cary 14 spectrophotometer was used in conjunction with the
electronics mentioned above. The reference electrode was a saturated calomel electrode (H-cell type) with saturated sodium
chloride replacing potassium chloride to avoid precipitation of
potassium perchlorate at the junction. A polyethylene tube with a
Vycor ("thirsty glass") plug and containing 0.5 M tetraethylammonium perchlorate (TEAP) in DMF served as a salt bridge connecting
the reference electrode and sample solution. A platinum wire was
used as the auxiliary electrode. All sample solutions (1-3 mM)
were thoroughly degassed (in a rubber stoppered container) with
argon that was first passed over hot copper to remove trace oxygen
and then through a gas scrubbing tower containing DMF. The Cary
14 cell compartment was modified to allow electrical leads to con-
nect the electrodes with the potentiostat and also was sealed with
clear polyethylene film before purging by argon. Argon was kept
flowing through the compartment for the duration of the experiment.
About 5 ml of sample solution was transferred by argon pressure
via a Teflon tube immersed in the solution and leading to the
sample cup containing the OTTLE. Capillary action caused the solution to fill the cell and aspiration was used to change solution
when necessary.

     <u>Materials</u>. The Schiff base ligands and metal complexes were
prepared and purified by established literature procedures (25).
Polarographic grade TEAP was purchased from G. Frederick Smith
Chemical Co. and in most cases was used without further purification; in one case this material was recrystallized from methanol,
and the resulting product gave results identical to those obtained
with the unpurified material. Fisher Certified DMF was used without further purification and also after purification by a literature
procedure (32); these two sources yielded identical experimental
results, in agreement with previous observations (33).

3. RESULTS AND DISCUSSION

The potential range of the Au minigrid OTTLE in DMF with 0.5 M TEAP is shown by voltammogram A in Figure 2. The usable potential range extends from +1.0 V to -1.6 V vs. SCE. The negative potential range could be extended by 0.4 V by coating the gold minigrid with a thin film of mercury as shown by voltammogram B.

Bis(salicylaldehyde)ethylenediiminecobalt (II). Co(II)sal$_2$en. Each compound studied is initially characterized by cyclic voltammetry in the OTTLE to locate the redox potentials for reduction/ oxidation of the parent compound and to evaluate the stability of the electrogenerated redox state(s). A cyclic voltammogram for Co(II)sal$_2$en reveals that the Co(II) is oxidized by a positive scan from $E_{initial}$ = -0.3 V and reduced by a negative scan as shown in Figure 3. The two redox couples are assigned as Co(II) $\rightleftarrows$ Co(III) with $E^{°\prime}$ = +0.040 V vs. SCE and Co(II) $\rightleftarrows$ Co(I) with $E^{°\prime}$ = -1.193 V vs. SCE. These $E^{°\prime}$ values are calculated by averaging the cathodic and anodic peak potentials. In principle, these peak potentials should be identical for a reversible couple in thin-layer cyclic voltammetry (34); however, the peaks are skewed by the iR drop which is rather large in the thin layer cell, especially in a nonaqueous solvent. The large resistance of the thin layer cell necessitates the use of rather slow scan rates for

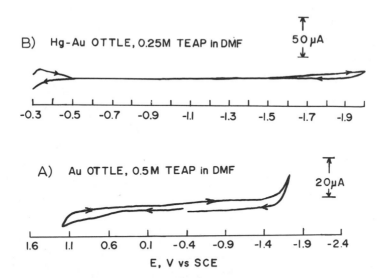

Figure 2. Cyclic voltammograms of supporting electrolyte/ solvent. (A) Au OTTLE, scan rate 2.0 mV sec^{-1}, initial potential -0.4 V vs. SCE. (B) Hg-Au OTTLE, scan rate 2.0 mV sec^{-1}, initial potential -0.3 V vs. SCE.

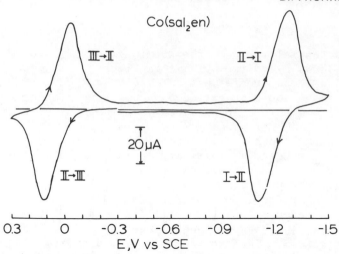

Figure 3.   Cyclic voltammogram of 1.3 mM Co(sal$_2$en) in 0.5M TEAP/DMF on Au OTTLE.   Scan rate 2.0 mV sec^{-1}, initial potential -0.3 V vs. SCE.

voltammetry.   The rapid drop in current after a peak coincides with the complete electrolysis of the thin solution layer.   The redox couples could be cycled indefinitely among the three redox states, indicating electrochemical reversibility and stability of the electrogenerated redox states on the slow time scale of the voltammogram (1-2 mV sec^{-1} scan rates).

The assignment of redox states to the electrogenerated species was confirmed by thin-layer controlled potential coulometry. Figure 4 shows a typical thin-layer charge-time curve for oxidation of Co(II) by a potential step from -0.3 V to +0.3 V vs. SCE.   The charge increases rapidly until the Co(II) in the thin layer is exhaustively oxidized at which time the curve levels off.   Also shown is a charge-time curve for supporting electrolyte alone. The n value for Co(II) oxidation is determined by Faraday's Law from the Faradaic charge, Q_F

$$Q_F = Q_T - Q_B = nFVC$$

where Q is charge in coulombs; n is the number of electrons transferred per molecule; F is Faraday's number, 96,493 coulombs per equivalent; V is the solution volume of the thin layer cell in liters; and C is the concentration of electroactive species in moles/liter.

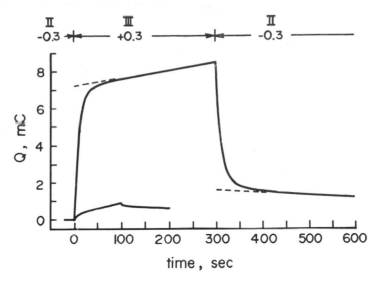

Figure 4.   Thin layer controlled potential coulometry of 2.1 mM Co(sal$_2$en) in 0.5 M TEAP/DMF.   Charge-time curve for potential step -0.3 V to +0.3 V to -0.3 V vs. SCE.

Calibration of the cell volume, V, is accomplished via a separate coulometry experiment with a standard ferricyanide solution.   Values for the total charge, Q_T, and the background charge, Q_B, for Co(sal$_2$en) were obtained by extrapolating to zero time as shown in the Figure.   Figure 4 also shows the effect of stepping the potential back to -0.3 V, causing reduction to the original Co(II) species.   Results for both the forward and reverse steps indicate a one-electron process: Co(II) → Co(III), n=0.95; Co(III) → Co(II), n=0.94.   Similar experiments on the Co(II) reduction wave also confirm a one electron process generating Co(I):   Co(II) → Co(I), n=1.03; Co(I) → Co(II), n=0.94.

    A spectropotentiostatic technique employing the OTTLE can be used in conjunction with the Nernst equation as a separate means for determining the values of E°' and n (20,35).   In an electrochemical cell the ratio of concentrations of oxidized to reduced forms of the electroactive species at the electrode surface is determined by the potential applied to the electrode as defined by the Nernst equation:

$$E_{applied} = E°' + \frac{0.059}{n} \log \frac{[0]}{[R]}$$

For the OTTLE, the potential applied to the minigrid determines this concentration ratio in the entire solution comprising the thin

layer cell. When $E_{applied}$ is changed, the ratio, [O]/[R], rapidly
adjusts by electrolysis to the Nernstian value, which can then be
observed spectrophotometrically.

Figures 5 and 6 show spectra of the cobalt compound for a
series of applied potentials. The absorbances at the various
wavelengths where maxima exist reflect the amount of cobalt in
oxidized/reduced forms via Beer's Law; thus the ratio [O]/[R] corresponding to each value of $E_{applied}$ can be calculated from the
spectra by the following equation:

$$[O]/[R] = \frac{A_2-A_1}{A_3-A_2}$$

where A_3 is the absorbance of the completely oxidized form, A_1 is
the absorbance of the completely reduced form, and A_2 is the
absorbance for the mixtures of oxidized and reduced forms.   An
absorbance versus applied potential plot as shown in Figure 7
summarizes the spectral changes which accompany the variation in

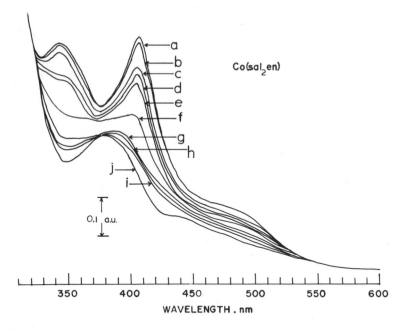

Figure 5.   Spectra of 1.3 mM Co(sal$_2$en) from spectropotentiostatic experiment.   Co(II) → Co(III) process. $E_{applied}$:   (a)
-0.300, (b) -0.050, (c) -0.003, (d) +0.010, (e) +0.020, (f) +0.040,
(g) +0.060, (h) +0.080, (i) +0.100, (j) +0.140 V vs. SCE.

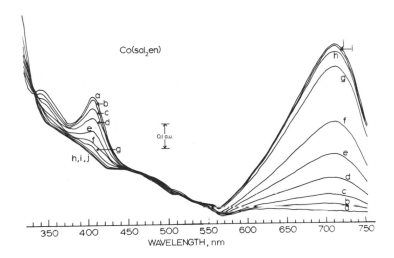

Figure 6.   Spectra of 1.3 mM Co(sal$_2$en) from spectropotentiostatic experiment.   Co(II) → Co(I) process. E$_{applied}$:   (a) -0.900,
(b) -1.120, (c) -1.140, (d) -1.160, (e) -1.180, (f) -1.200, (g)
-1.250, (h) -1.300, (i) -1.400, (j) -1.450 V vs. SCE.

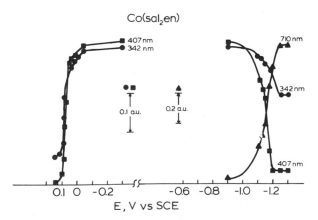

Figure 7.   Absorbance vs. potential plot for Co(sal$_2$en),
Co(II) → Co(III) and Co(II) → Co(I) processes, from spectra of
Figures 5 and 6.

redox state of Co(sal$_2$en). Figure 8 shows a plot of E$_{applied}$ versus log [O]/[R] for the data at 342 and 407 nm for both Co(II) → Co(III) and Co(II) → Co(I). The plots are linear as predicted by the Nernst equation and yield n values of 0.93 for the II, III process and 1.0 for the II, I couple. The intercepts give E°' values of +0.029 V and -1.193 V, respectively.

It is interesting to note that both the Co(II,I) and Co(II, III) redox couples exhibit thermodynamically reversible behavior in the OTTLE spectropotentiostatic experiment, the Nernstian plots in Figure 8 being linear and of the proper slope. By contrast these couples exhibit "quasi-reversible" behavior under the conditions of conventional (non-thin layer) cyclic voltammetry. For example, a cyclic voltammogram on a platinum wire electrode at a scan rate of 20 mV sec^{-1} gives a ΔE_p of 70 mV for the Co(II,I) couple and 115 mV for the Co(II,III) couple, values which are greater than the expected ΔE_p of 59/n mV. Thus, although these systems are quasi-reversible under such conditions, they are reversible under the very slow time scale of the OTTLE spectropotentiostatic experiment, where each potential is maintained until equilibrium is reached; and thermodynamically meaningful values of E°' can be obtained from the Nernstian plot. These

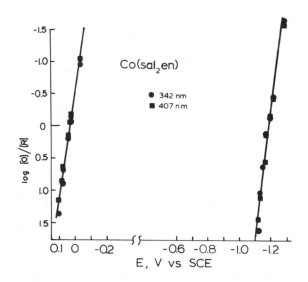

Figure 8. Plot of E$_{applied}$ vs. log [O]/[R] for Co(sal$_2$en), Co(II) → Co(III) and Co(II) → Co(I) processes.

results suggest that the OTTLE technique may be generally applicable for the measurement of reduction potentials of the many redox systems which fall into this category of "quasi-reversibility".

The spectra of the Co(II) and Co(I) species shown in Figure 6 also demonstrate one significant advantage of the OTTLE technique over more conventional procedures for obtaining spectra of air sensitive species. While the Co(II) spectrum is in good agreement with previously published results (36,37), it is clear from Figure 6 that previously published spectra (36,37) of the Co(I) species were actually obtained using solutions that contained significant amounts of the Co(II) species (e.g. the reported "Co(I)" peak at ca. 350 nm is seen to arise from the Co(II) peak at 342 nm; the reported absorptivities for the Co(I) peak at 710 nm are significantly lower than our value since Co(II) does not absorb at this wavelength). In the OTTLE, exhaustive electrolysis insures that only the Co(I) species is present in the light beam; with conventional techniques it is very difficult to manipulate Co(I) solutions without suffering some extraneous oxidation to Co(II). The low energy, intense (ε = 14,000 M^{-1}cm^{-1}) Co(I) peak at 710 nm is metal-to-ligand charge transfer band (38,39).

Bis(salicylaldehyde)ethylenediiminecopper(II). As before, cyclic voltammetry is employed initially to electrochemically characterize the complex. Figure 9 shows that this compound is reduced by a negative scan initiated at -0.7 V. This couple is assigned as Cu(II) → Cu(I) with $E^{\circ\prime}$ = -1.150 V vs. SCE, the $E^{\circ\prime}$ value being determined in the same fashion as for the cobalt analog. Again, thin layer controlled potential coulometry confirms this assignment. A potential step of -0.7 V to -1.4 V yields n=1.01. Spectropotentiostatic results which are summarized in Figures 10, 11 and 12, give $E^{\circ\prime}$ = -1.164 V vs. SCE and n = 1.09.

4.   CONCLUSIONS

Application of the OTTLE techniques in a nonaqueous solvent system has proven to be efficacious in spectrally and electrochemically characterizing the various oxidation states of two typical Schiff base complexes (i.e. the cobalt and copper complexes of bis (salicylaldehyde)ethylenediimine). Spectropotentiostatically obtained n values compare favorable with those obtained by thin layer controlled potential coulometry and with the values predicted from the known chemistries of copper and cobalt complexes. $E^{\circ\prime}$ values estimated by cyclic voltammetry are in good agreement with those obtained by Nernstian analysis in this work facilitate the acquisition of two types of data which are not readily available

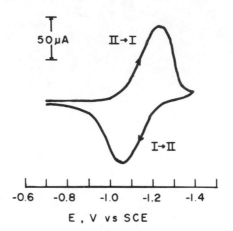

Figure 9.   Cyclic voltammogram of 2.2 mM Cu(sal$_2$en) in 0.5 M TEAP/DMF on Au OTTLE.   Scan rate 2.0 mV sec^{-1}, initial potential -0.7 V vs. SCE.

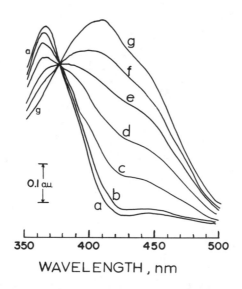

Figure 10.   Spectra of 2.2 mM Cu(sal$_2$en) from spectropotentio-static experiment, Cu(II) → Cu(I) process.   E$_{applied}$:   (a) -0.700, (b) -1.000, (c) -1.050, (d) -1.100, (e) -1.125, (f) -1.175, (g) -1.300 V vs. SCE.

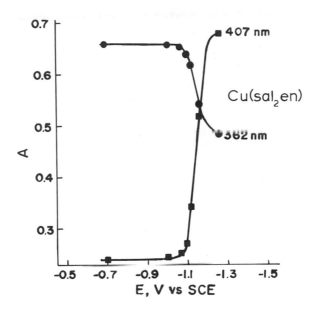

Figure 11.   Absorbance vs. potential plot for Cu(sal$_2$en), Cu(II) → Cu(I) process, from spectra of Figure 10.

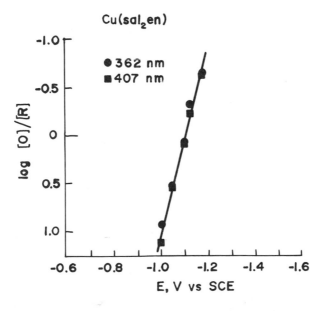

Figure 12.   Plot of $E_{applied}$ vs. log [0]/[R] for Cu(sal$_2$en), Cu(II) → Cu(I) process.

from more conventional techniques.   (1) Exhaustive electrolysis of
the small volume of solution contained in the OTTLE insures the
quantitative generation of a particular redox state without interference from extraneous redox reactions.   Thus, for example, in
this work we have obtained the spectrum of a very air sensitive
cobalt(I) species without the cobalt(II) interference which has
plagued earlier studies.   (2) In this work we show that whereas
cyclic voltammetry yields only quasi-reversible waves and therefore
the magnitude of the peak separation does not yield accurate n
values, the OTTLE spectropotentiostatic technique does yield Nernstian potentials and accurate n values.

5.   SUMMARY

The usefulness of thin layer   spectroelectrochemical techniques
for the characterization of Schiff base metal complexes in a nonaqueous medium is evaluated.   For two specific compounds chosen as
model systems, i.e. the cobalt and copper complexes of bis(salicylaldehyde)ethylenediimine, in N,N-dimethylformamide with tetraethylammonium perchlorate as supporting electrolyte, these techniques
efficaciously permit determination of redox potentials, number of
electrons transferred in the redox process, and spectra of the
various oxidation states of the complexes.   The gold minigrid
OTTLE provides a potential window which is suitable for the investigation of these complexes and which would seem to be suitable
for the investigation of many Schiff base complexes; the cathodic
limit of this window may be further extended by mercury deposition
on the gold surface.   The utility of the OTTLE techniques demonstrated in this work encourages us to apply these techniques to a
wide variety of structurally modified Schiff base complexes of
various metals in an attempt to correlate the spectral and electrochemical properties of these complexes.   The results of these
extended studies will be the subject of future reports.

6.   ACKNOWLEDGMENTS

We thank the research group of Dr. W. R. Heineman for helpful
consultations, especially Barbara J. Norris and Thomas P. DeAngelis
with regard to experimental technique and Michael L. Meyer with
regard to the mercury coated gold OTTLE.   This research was partially supported by the National Science Foundation, Grant
GP-41981X.

7.   REFERENCES

(1)   R. W. Murray, W. R. Heineman, and G. W. O'Dom, Anal. Chem.,
      39, 1666 (1967).

(2)   W. R. Heineman, J. N. Burnett, and R. W. Murray, Anal. Chem.,
      40, 1970 (1968).

(3)   W. R. Heineman, J. N. Burnett, and R. W. Murray, Anal. Chem.,
      40, 1974 (1968).

(4)   A. Yildiz, P. T. Kissinger, and C. N. Reilley, Anal. Chem.,
      40, 1018 (1968).

(5)   I. Piljac and R. W. Murray, J. Electrochem. Soc., 118, 1758
      (1971).

(6)   R. M. Wightman, J. R. Cockrell, R. W. Murray, J. N. Burnett,
      and S. B. Jones, J. Amer. Chem. Soc., 98, 2562 (1976).

(7)   T. E. Neal and R. W. Murray, Anal. Chem., 42, 1654 (1970).

(8)   G. Peychal-Heiling and G. S. Wilson, Anal. Chem., 43, 545,
      550 (1971).

(9)   B. P. Neri and G. S. Wilson, Anal. Chem., 44, 1002 (1972);
      45, 442 (1973).

(10) V. S. Srinivasan and F. C. Anson, J. Electrochem. Soc., 120,
      1359 (1973).

(11) T. M. Kenyhercz, T. P. DeAngelis, B. J. Norris, W. R.
      Heineman, and H. B. Mark, Jr., J. Amer. Chem. Soc., 98, 2469
      (1976).

(12) T. M. Kenyhercz, H. B. Mark, Jr., and P. T. Kissinger,
      ACS Symposium Series, in press.

(13) W. R. Heineman, B. J. Norris, and J. F. Goelz, Anal. Chem.,
      47, 79 (1975).

(14) F. M. Hawkridge and B. Ke, Anal. Biochem., in press.

(15) E. A. Blubaugh, A. M. Yacynych and W. R. Heineman, unpublished
      results.

(16) R. L. McCreery, Anal. Chem., submitted for publication.

(17)   C. A. Owens and G. L. Dryhurst, J. Electroanal. Chem., in
       press.

(18)   W. R. Heineman, T. P. DeAngelis, and J. F. Goelz, Anal.
       Chem., 47, 1364 (1975).

(19)   M. L. Meyer, T. P. DeAngelis, and W. R. Heineman, Anal. Chem.,
       in press.

(20)   B. J. Norris, M. L. Meckstroth, and W. R. Heineman, Anal.
       Chem., 48, 630 (1976).

(21)   I. Piljac, M. Tkalcec, and B. Grabaric, Anal. Chem., 47,
       1369 (1975).

(22)   P. T. Kissinger and C. N. Reilley, Anal Chem., 42, 12 (1970).

(23)   A. Bigotto, G. Costa, G. Mestroni, G. Pellizer, A. Puxeddu,
       E. Reisenhofer, L. Stefani, and G. Tauzher, Inorg. Chim.
       Acta Revs., 4, 41 (1970).

(24)   G. S. Patterson and R. H. Holm, Bioinorg. Chem., 4, 257
       (1975).

(25)   R. H. Holm, G. W. Everett, Jr., and A. Chakravorty, Prog.
       Inorg. Chem., 7, 83 (1966).

(26)   M. Calvin and R. H. Bailes, J. Amer. Chem. Soc., 68, 949
       (1946).

(27)   J. R. Urwin and B. O. West, J. Chem. Soc., 4727 (1952).

(28)   M. J. Carter, L. M. Engelhardt, D. D. Rillema and F. Basolo,
       J. Chem. Soc. Chem. Commun., 810 (1973).

(29)   R. P. Hanzlik and D. F. Smith, Jr., J. Chem. Soc. Chem.
       Commun., 528 (1974).

(30)   T. M. Krygowski and W. R. Fawcett, J. Amer. Chem Soc., 97
       2143 (1975).

(31)   W. R. Fawcett and T. M. Krygowski, Aust. J. Chem., 28, 2115
       (1975).

(32)   P. Kanatharan and M. S. Spritzer, Anal. Lett., 6, 421 (1973).

(33)   C. K. Mann in "Electroanalytical Chemistry," Vol. 3, A. J.
       Bard, ed., Marcel Dekker, New York, 1969, p. 79.

(34)   A. T. Hubbard and F. C. Anson in "Electroanalytical Chemis-
       try," Vol. 4, A. J. Bard, ed., Marcel Dekker, New York, 1970.

(35)   T. P. DeAngelis and W. R. Heineman, J. Chem. Ed., 53, 594
       (1976).

(36)   G. Costa, G. Mestroni and G. Pellizer, J. Organomet. Chem.,
       11, 333 (1968).

(37)   F. Calderazzo and C. Floriani, Chem. Commun., 139 (1967).

(38)   P. K. Das, H.A.O. Hill, J. M. Pratt and R.J.P. Williams,
       Biochim. Biophys. Acta, 161, 646 (1967).

(39)   P. Day, G. Schregg and R.J.P. Williams, Biopolymers Symp.,
       1, 271 (1964).

ELECTROCHEMICAL AND SPECTROSCOPIC STUDIES OF RHODIUM AND IRIDIUM DIIMINE COMPLEXES IN ACETONITRILE

Kenneth W. Hanck, M. Keith DeArmond, Gregory Kew,
James L. Kahl and Horia Căldăraru†
Department of Chemistry, North Carolina State
University
Raleigh, North Carolina 27607

1.   INTRODUCTION

The diimine and triimine complexes of the first row transition metals have been examined extensively in non-aqueous solvents (1-13). A number of these studies have employed both spectroscopic and electrochemical techniques to characterize the species produced during electrolysis. In many cases unusual species result when electrons are transferred to diimine chelates; the added electron delocalizes onto the ligand and the oxidation state of the metal remains unchanged. The nominal d^6 complexes of Group VIII which have been studied and the species produced from them are summarized in Table I.

The platinum group metals are known to form diimine complexes. We sought to determine the nature of the species produced by the addition of electron(s) to the d^6 starting material. Experiments with the bis and tris complexes of 2,2'-bipyridyl (bpy) and 1,10-phenanthroline (phen) with Rh(III), Ir(III) and Ru(II) will be described.

The objectives of our research were four fold: (1) establish the electron transfer sequence for each species, (2) determine the chemical formula of the electrolytic products, (3) determine the type of orbitals involved in the charge transfer reaction, and (4) establish the molecular geometry of any delocalized species produced.

----

†Visiting Scientist from the Quantum Chemistry Department, Institute of Physical Chemistry, Bucharest 9, Romania.

TABLE I

Species Generated by Electrochemical Techniques
From d^6 Chelates

Ligand	$[FeL_3]^{2+}$	$[FeL_3]^{1+}$	$[FeL_3]^0$	$[FeL_3]^{1+}$
bpy	ℓ	d	d	d
phen	ℓ	d	d	d
4,7-dimethyl phen	ℓ	d	d	d
4,7-diphenyl phen	ℓ	d	d	d
2,2',2"-terpyridine	ℓ	d	d	-
	$[CoL_3]^{3+}$	$[CoL_3]^{2+}$	$[CoL_3]^{1+}$	$[CoL_3]^{1-}$
bpy	ℓ	ℓ	d	d

ℓ = species in which redox orbital is a localized metal
        orbital
d = species in which redox orbital is a delocalized orbital

Clearly a single method cannot be relied upon to achieve
these objectives. Consequently, we have utilized cyclic voltammetry, pulse voltammetry, controlled potential electrolysis,
luminescence spectroscopy and electron spin resonance spectroscopy
in our investigations.

All of the d^6 complexes studied luminesce in the visible region.
The pertinent luminescence parameters are summarized in Table II.
Although beyond the scope of this chapter the nature of the lowest
empty orbital of the emitting species can be deduced from the data
of Table II (14,15). As a first approximation the orbital form
which emission occurs can be assumed to be the orbital into which
the first electron will be transferred when the complex is reduced
electrochemically. The wave function for the lowest empty orbital
is in all cases most precisely described by a linear combination
of metal d orbitals and ligand π orbitals with the relative magnitudes of metal and ligand orbital mixing varying considerably for
the various species.

TABLE II

Luminescence Parameters for Nominal d^6 Complexes of
Ru^{2+}, Os^{2+}, Rh^{3+} and Ir^{3+} with Diimine Ligands

Complex	$\bar{\nu}_{max}$ (x 10^{-3})	τ_p (77°K)	Φ_p (77°K)	Band Shape	Stokes Shift	Emitting Orbital Character
$[Rh(bpy)_3]^{3+}$	22.2	2.21	0.30	structured	small	d
$[Rh(phen)_3]^{3+}$	22.2	48.4	1.0	structured	small	d
$[Rh(bpy)_2(phen)]^{2+}$	22.2	37 and 2.3	0.7-0.6[a]	structured	small	d
$[Rh(bpy)(phen)_2]^{3+}$	22.3	36 and 2.3	0.6-0.3[a]	structured	small	d
$[Rh(bpy)_2Cl_2]^+$	14.2	47.3 x 10^{-3}	0.02	structureless	large	ℓ
$[Rh(phen)_2Cl_2]^+$	14.1	41.5 x 10^{-3}	0.025	structureless	large	ℓ
$[Ru(bpy)_3]^{2+}$	17.3[b]	5.2 x 10^{-3}	0.38	structured	small	d
$[Ru(phen)_3]^{2+}$	17.7[b]	9.8 x 10^{-3}	0.58	structured	small	d
$[Ir(bpy)_2Cl_2]^+$	21.3	5.4 x 10^{-3}	0.44	structured	small	d
$[Ir(phen)_2Cl_2]^+$	21.0	7.3 x 10^{-3}	0.35	structured	small	d
$[Os(bpy)_3]^{2+}$	17.4[c]	0.9 x 10^{-3}	0.03	structured	small	d
$[Os(phen)_3]^{2+}$	16.7	2.43 x 10^{-3}	0.13	structured	small	d

$\bar{\nu}_{max}$ is the energy in cm^{-1} of emission maxima

τ_p is the experimental lifetime in millisec. of the emission

Φ_p is the emission quantum yield in alcohol solvents

a = varies with excitation wavelength; b = G.A. Crosby, W.G. Perkins and D.M. Klassen, J. Chem. Phys., 43, 1498 (1965).; c = F. Zuloaga and M. Kasha, Photochem. & Photobiol., 7, 549 (1968).

$$\psi_{emiss} = a\phi_M + b\phi_2 \qquad\qquad (1)$$

In most cases one orbital source or the other predominates, i.e., the emitting orbital is a localized (d) orbital (a > b) or a de-localized (π ligand) type (b > a). Table II summarizes the con-clusions concerning the nature of the lowest empty orbital in diimine complexes.

A complex in which the lowest empty orbital has considerable ligand character would be more likely to produce a stable delo-calized species upon electroreduction than one in which the lowest empty orbital has considerable metal character. Our expectation was that $[RhL_3]^{3+}$ would be the most likely to yield a stable delocalized species upon addition of one electron while $[RhL_2Cl_2]^{1+}$ would be the least likely. The complexes $[IrL_2Cl_2]^{1+}$ and $[RuL_3]^{2+}$ might be expected to be of intermediate character with the Ir com-plex perhaps expected to produce a more stable species than the Ru complex because of its greater ligand character in the lowest empty orbital.

2. EXPERIMENTAL

All measurements were performed in acetonitrile containing 0.1 M tetraethylammonium perchlorate; the purification of both materials has been described (16). The complexes were synthesized using literature procedures (14,17). The electrochemical (16,18), luminescence (14) and electron spin resonance instrumentation (19) have been described elsewhere.

3. RESULTS

Tris complexes of Rh(III). The cyclic voltammograms of $[Rh(bpy)_3]^{3+}$ indicate the presence of 4 reduction processes (Figure 1). At slow sweep rates 3 cathodic and 2 anodic peaks are observed. The peak at ca. - 0.8V is higher than either of the subsequent cathodic peaks; the peak current is less than that ex-pected for a two electron process. At fast sweep rates the peak at ca. -0.8V splits into 2 distinct peaks while the peaks more negative than -1.1V remain unaffected.

A pale-yellow colored solution is produced during controlled potential electrolyses of $[Rh(bpy)_3]^{3+}$ solutions at -1.1V. Lumi-nescence spectra of the electrolyzed solution indicate the presence of free bipyridyl in amounts consistent with the loss of 1 bpy ligand per molecule of complex. The species $[Rh(bpy)_2]^{1+}$ is ex-pected after the addition of 2 electrons and the loss of 1 bpy. Martin, McWhinnie and Waind have synthesized and characterized $[Rh(bpy)_2]ClO_4 \cdot 3H_2O$ (17). This material when dehydrated and dissolved in acetonitrile yielded the voltammogram shown in Figure

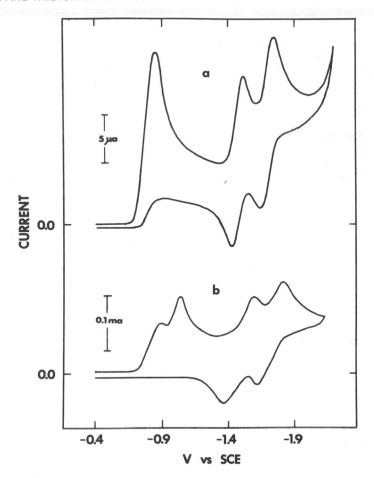

Figure 1.   Cyclic voltammograms of $1.08 \times 10^{-4}\underline{M}$ $[Rh(bpy)_3]^{3+}$;
(a) 0.10 V/sec, (b) 31.2 V/sec.

2.   Clearly the peaks in Figure 1 more cathodic than -1.1V agree
with the 2 peaks of Figure 2.

    The cyclic voltammetry of $[Rh(phen)_3]^{3+}$ is similar to that of
$[Rh(bpy)_3]^{3+}$ (Figure 3).   There are 2 important differences, however.   First the 2 peaks in the vicinity of -0.8 V appear at slow
sweep rates; as shown in Figure 3a the product of the first reduction step can be reoxidized at a sweep rate of 0.10 V/sec while
46.1 V/sec were required for the corresponding bpy complex (16).

    The second feature which appears is adsorption.   Cyclic
voltammograms taken with freshly cleaned electrodes show several

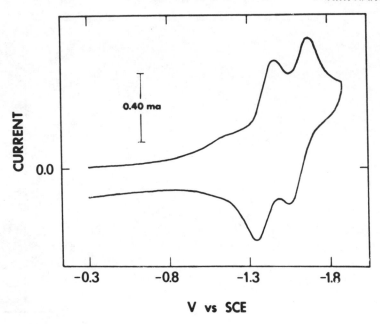

Figure 2.   Cyclic voltammogram of $5.15 \times 10^{-4}\underline{M}$ $[Rh(bpy)_2]^{1+}$ at 9.65 V/sec.

spike-like adsorption peaks. We have previously discussed these phenomena; they may be eliminated by proper pre-treatment of the electrode (18).

Changing the ligand from bpy to phen does not change the electrode mechanism and, in fact, barely alters the peak potential of each reduction step (Table III). The phen ligand is larger and capable of greater electron delocalization; consequently the species produced by electrolysis of $[Rh(phen)_3]^{3+}$ are more stable than their bpy counterparts.

Two mixed ligand complexes have been synthesized in our laboratory (20). The voltammetry of these materials is nearly identical to that of $[Rh(phen)_3]^{3+}$ (Figure 4). The luminescence decay of the mixed ligand complexes is best fit by a sum of two exponentials with one decay constant consistent with that of $[Rh(bpy)_3]^{3+}$ and the other nearly identical to that of $[Rh(phen)_3]^{3+}$. Consequently the small magnitude of the d orbital mixing coeffi- cient in these tris complexes precludes conjugation through the metal atom; each ligand acts independently of the other two. Phenanthroline is believed to weakly adsorb on the Pt electrode through the 5 - 6 bond, leading to the preferential transfer of

TABLE III

Peak Potentials* of Rh bpy/phen Complexes

	(E_{pc}) I	(E_{pa}) I	(E_{pc}) II	(E_{pa}) II	(E_{pc}) III	(E_{pa}) III	(E_{pc}) IV	(E_{pa}) IV
$[\text{Rh(phen)}_3]^{3+}$	-0.79	-0.73	-0.93	---	-1.45	-1.39	-1.63	-1.58
$[\text{Rh(phen)}_2(\text{bpy})]^{3+}$	-0.79	-0.71	-0.93	---	-1.44	-1.38	-1.65	-1.59
$[\text{Rh(phen)(bpy)}_2]^{3+}$	-0.79	-0.72	-0.94	---	-1.47	-1.41	-1.68	-1.62
$[\text{Rh(bpy)}_3]^{3+}$	-0.83	(I + II)	---	---	-1.46	-1.40	-1.67	-1.61
$[\text{Rh(phen)}_2\text{Cl}_2]^{1+}$	-0.83	(I + II)			-1.45	-1.39	-1.67	-1.61
$[\text{Rh(bpy)}_2\text{Cl}_2]^{1+}$	-0.84	(I + II)	---	---	-1.46	-1.40	-1.67	-1.61

*E_p in V vs aq. SCE; v = 0.1 V/sec

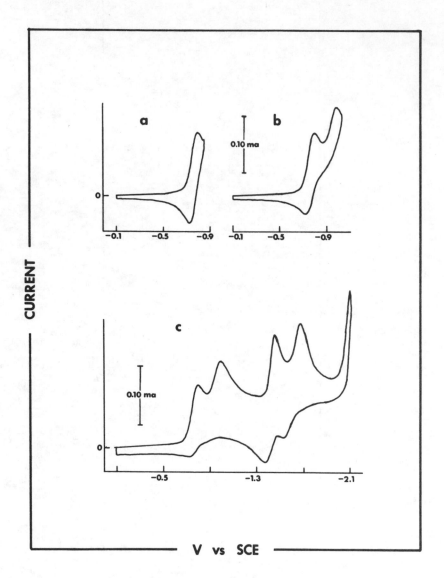

Figure 3.   Cyclic voltammograms of 4.90 x $10^{-4}\underline{M}$ [Rh(phen)$_3$]$^{3+}$ at 0.10 V/sec.

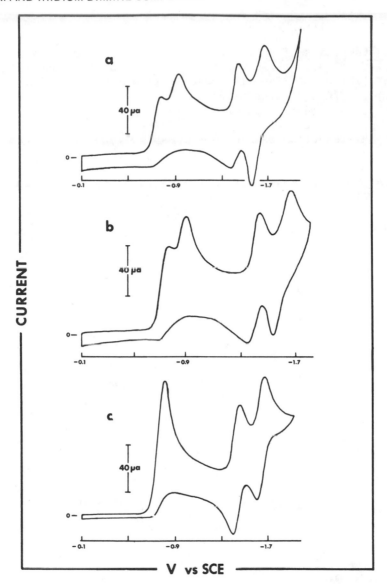

Figure 4.   Cyclic voltammograms of mixed ligand complexes at 0.10 V/sec.

(a)   4.53 x 10^{-4}M $[Rh(phen)_2(bpy)]^{3+}$
(b)   5.47 x 10^{-4}M $[Rh(bpy)_2(phen)]^{3+}$
(c)   6.27 x 10^{-4}M̲ $[Rh(bpy)_3]^{3+}$

the first electron through a phen ligand. Our data are not precise enough to determine whether the peak potential is dependent on the phen to bpy ratio; clearly the morphology of the voltammograms of

$[\text{Rh}(\text{phen})_3]^{3+}$, $[\text{Rh}(\text{phen})(\text{bpy})_2]^{3+}$ and $[\text{Rh}(\text{phen})_2\text{bpy}]^{3+}$ is identical.

    <u>Bis Complexes of Rh(III)</u>. The cyclic voltammograms of $[\text{Rh}(\text{bpy})_2\text{Cl}_2]^+$ are nearly invariant with sweep rate and resemble those of $[\text{Rh}(\text{bpy})_3]^{3+}$ at slow sweep rate (Figure 5). Solutions of $[\text{Rh}(\text{bpy})_2\text{Cl}_2][\text{ClO}_4]$ after electrolysis at -1.1V do not contain free bpy but do contain free chloride. The peaks more negative than - 1.1V agree closely with those of $[\text{Rh}(\text{bpy})_2]^{1+}$ (see Figure 2). The initial reduction peak cannot be split into two 1 electron peaks with increasing sweep rate as for $[\text{Rh}(\text{bpy})_3]^{3+}$.

    The electrochemistry of $[\text{Rh}(\text{phen})_2\text{Cl}_2]^{1+}$ parallels that of $[\text{Rh}(\text{bpy})_2\text{Cl}_2]^{1+}$ (Figure 6). The contrast between the voltammetry

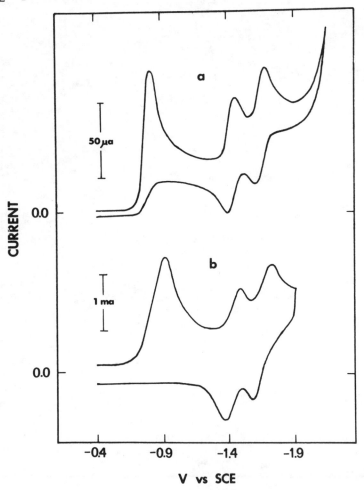

Figure 5. Cyclic voltammogram of 5.16 x $10^{-4}\underline{\text{M}}$ $[\text{Rh}(\text{bpy})_2\text{Cl}_2]^{1+}$;

(a) 0.10 V/sec, (b) 32.2 V/sec.

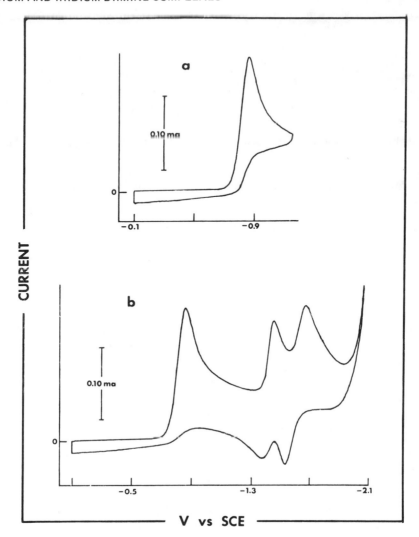

Figure 6. Cyclic voltammograms of a saturated solution of
Rh(phen)$_2$Cl$_2$ $^{1+}$ at 0.10 V/sec.

of [Rh(phen)$_3$]$^{3+}$ and [Rh(phen)$_2$Cl$_2$]$^{1+}$ is striking. The tris
species shows 4 distinct processes, the bis only 3; the product of
the first reduction step may be reoxidized for the tris complex
but not for the bis; the voltammetry more negative than -1.1V is,
however, identical for the two complexes.

Electron transfer sequence for Rh(III) complexes. The
reactions taking place during the reduction of bpy and phen complexes of Rh(III) are summarized in Figure 7. Although there are

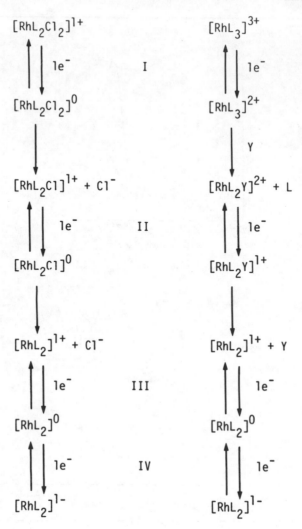

Figure 7. Summary of electron transfer sequence for
$[Rh(L)_3]^{3+}$ and $[RhL_2Cl_2]^{1+}$.

distinct differences in the voltammetry of bpy and phen complexes,
complexes of either ligand undergo the same series of electrode
reactions hence in Figure 7 the symbol L is used to represent
either ligand.

The single large wave observed at ca. -0.8V is in fact a
composite of two 1 electron processes each of which is followed by
a chemical reaction. The product of electron transfer I can be
reoxidized only for tris complexes; the product of electron
transfer II cannot be reoxidized for either tris or bis complexes

within the limits of our instrumentation. We infer from this that
the chemical reaction following the transfer of the second electron
is much faster than that following the first electron transfer.
As the scan rate is increased $(E_p)_{II}$ shifts cathodic at a greater
rate than $(E_p)_I$ thus "resolving" the peak at ca. -0.8V into two 1
electron peaks.

     Since the loss of a bidentate ligand requires that 2 coordi-
nate bonds be broken, loss of ligand is indicated in Figure 7 as
the chemical reaction following electron transfer I.   Rh(II)
species are expected to be at least penta coordinate (21), consequently a solvent molecule Y is believed to enter the coordination
sphere when the ligand is lost.   The loss of a monodentate ligand
such as chloride is expected to be very rapid, consequently the
chemical reactions following the first 2 electron transfers to the
bis complexes are much more rapid than those following the same
electron transfers to the tris complexes.   Since the last 2 voltammetric peaks are identical for bis and tris complexes, the
species after the transfer of 2 electrons must be the same regardless of whether the bis or tris complex is electrolyzed.

     The 2 most cathodic electron transfers are free of chemical
complications and exhibit "reversible behavior." The heterogeneous
charge transfer rate constants for electron transfers III and IV
have been measured from the separation of E_{pc} and E_{pa} (Table IV).
They are of the same order of magnitude as those reported for the
corresponding Os and Ru complexes (13).

TABLE IV

Heterogeneous Charge Transfer Rate Constants

	$(k_s)_{II}$	$(k_s)_{III}$	$(k_s)_{IV}$
$[Rh(phen)_3]^{3+}$	----	0.22 ± 0.05	0.15 ± 0.04
$[Rh(phen)_2(bpy)]^{3+}$	----	----	0.13 ± 0.04
$[Rh(phen)(bpy)_2]^{3+}$	----	----	0.18 ± 0.04
$[Rh(bpy)_3]^{3+}$	----	0.14 ± 0.04	0.08 ± 0.04
$[Ru(bpy)_3]^{2+}$	0.24*	0.26*	
$[Os(bpy)_3]^{2+}$	0.25*	0.18*	

*from (13)

Bis complexes of Ir(III). Our studies of these materials indicate that the voltammetry is dependent upon the purity of the complexes examined. The voltammetry described here was performed on samples recrystallized several times from acetonitrile. The impurities which complicate the voltammograms have not yet been identified but the preliminary luminescence measurements suggest that equilibria involving water chemically bound to a diimine ligand could be the source of the problem (22).

Cyclic voltammetry of $[Ir(bpy)_2Cl_2]^{1+}$ indicates 4 regions of electron activity (Figures 8 and 9). At ca. -1.2V a 1 electron process is observed which is free of chemical complications and exhibits "reversible behavior." At ca. -1.4V a second 1 electron process is observed which exhibits reversible behavior at sweep rates greater than 4 V/sec but is complicated by a following chemical reaction as evidenced by an increase in i_{pa}/i_{pc} with increasing sweep rate. The chemical reaction following the addition of the second electron appears to follow several pathways and produces species which can be further reduced. This is clearly shown in Figure 8. At ca. -2V a third electroreduction step is observed; the concentration of the species responsible for this process is dependent on the chemical steps between electron transfers II and III. At fast sweep rates $(i_{pc})_{III} \simeq (i_{pc})_{II} = (i_{pc})_I$ while at slow sweep rates $(i_{pc})_{III} \simeq 0$.

At ca. +2V a 1 electron oxidation is observed which exhibits "reversible behavior" at sweep rates greater than 20V/sec. The product of this oxidation decomposes before it can be re-reduced if slow sweep rates are used. The exact nature of the decomposition has not been determined but the process seems to be catalytic and involves oxidation of the solvent with consequent production of an oxidizable Ir(III) species since the anodic peak current increases beyond that expected for a 1 electron process as the sweep rate is lowered.

Voltammograms of $[Ir(phen)_2Cl_2]^{1+}$ are similar to those of $[Ir(bpy)_2Cl_2]^{1+}$. As in the case of Rh(III), the Ir(III) complexes follow the same basic electron transfer sequence regardless of which ligand is used. The phen complexes yield species which are more stable than the corresponding bpy species. This is especially true of the nominally Ir(IV) species produced at ca. +2V (Figure 10).

ESR studies of reduced Rh diimines. Several Rh diimine species have been examined with ESR at both room temperature and at 77°K (Table V). Only the $[RhL_2]^0$ species were paramagnetic (19); these species produced two lines (Figures 11 and 12).

The anisotropic line shape of the g ~ 2 signal is characteristic of an axially distorted S = 1/2 complex with $g_\perp > g_{||}$ (Table VI).

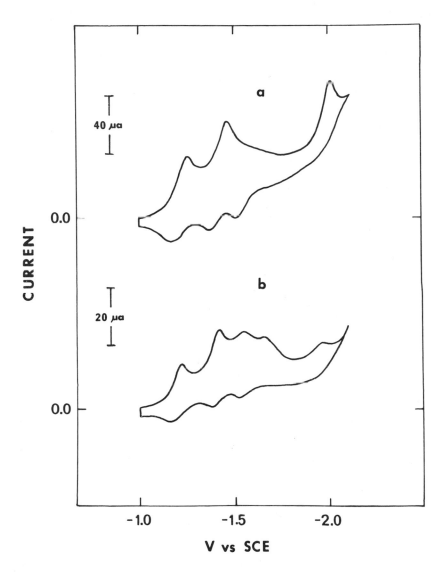

Figure 8.   Cyclic voltammograms of 6.0 x $10^{-4}\underline{M}$ $[Ir(bpy)_2Cl_2]^{1+}$;
(a)   4.45 V/sec,   (b)   0.460 V/sec.

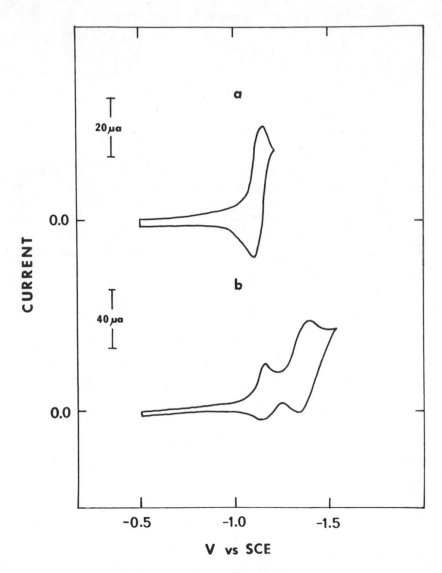

Figure 9. Effect of length of potential sweep on cyclic voltammograms of $6.0 \times 10^{-4}\underline{M}$ $[Ir(bpy)_2Cl_2]^{1+}$ at 0.460 V/sec.

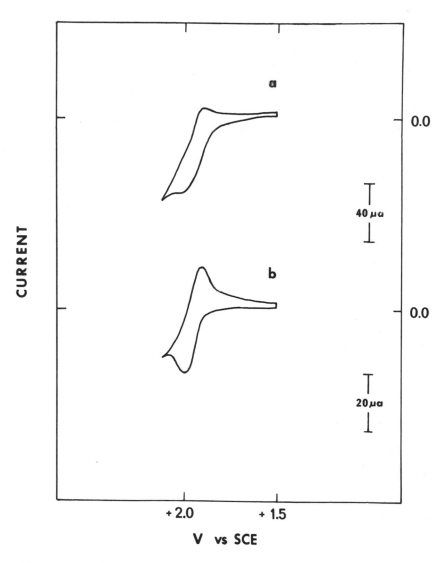

Figure 10.   Oxidation of Ir(III) diimines at 4.46 V/sec;

(a)   6.0 x $10^{-4}\underline{M}$ [Ir(bpy)$_2$Cl$_2$]$^{1+}$;
(b)   6.0 x $10^{-4}\underline{M}$ [Ir(phen)$_2$Cl$_2$]$^{1+}$

TABLE V

Species Examined With ESR

	Room Temp.	77°K
$[RhL_3]^{3+}$	NS	NS
$[RhL_2Cl_2]^{1+}$	NS	NS
$[RhL_2]^{1+}$	NS	NS
$[RhL_2]^{0}$	NS	$\{\!\!\downarrow \begin{array}{l} g \sim 2 \\ g \sim 4 \end{array}$

L = bpy or phen
NS = no signal

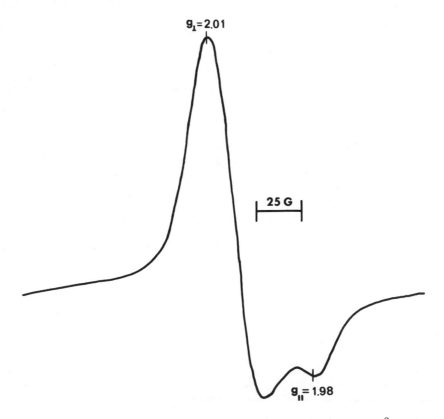

$g_\perp = 2.01$

25 G

$g_{\parallel} = 1.98$

Figure 11. Glassy solution ESR spectrum of $[RhL_2]^{0}$ in the g ∿ 2 region.

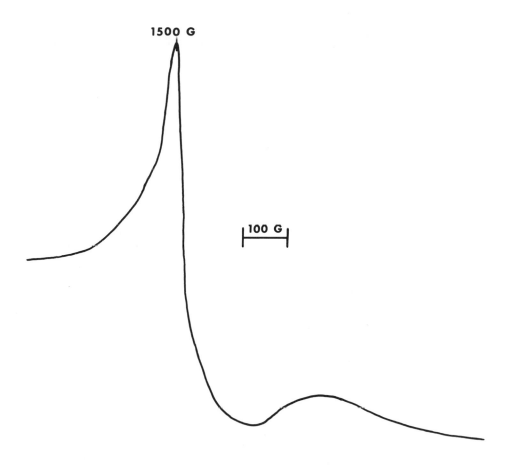

Figure 12. Glassy solution ESR spectrum of $[RhL_2]^0$ in the g $\sim$ 4 region.

TABLE VI

g Values for $[RhL_2]^0$ Complexes

	$g_{\parallel}$	$g_{\perp}$
$[Rh(bpy)_2]^0$	1.98 ± 0.003	2.01 ± 0.003
$[Rh(phen)_2]^0$	1.97 ± 0.003	2.01 ± 0.003

The sign of the g factor anisotropy and the absence of an ESR signal at room temperature are not typical of a d^9 complex. Furthermore the magnitude of the g anisotropy is much less than that of a typical d^9 complex; consequently, the $[RhL_2]^0$ species cannot be considered a d^9 crystal field complex. The complex cannot be considered as having the electron completely ligand localized to give an isolated bpy^{1-} or $phen^{1-}$ radical and Rh(I) since such a species would likely have an isotropic g factor (23). Therefore, $[RhL_2]^0$ is best described as a delocalized orbital complex, i.e., a metal containing heterocycle with the unpaired electron in a large delocalized orbital encompassing metal and the two diimine rings. Such an electronic structure accounts for the relative stability of $[RhL_2]^+$ and $[RhL_2]^0$ and is consistent with the distorted planar trans structure determined by X-ray diffraction for $[Pd(bpy)_2]^{2+}$ (24,25) which is isoelectronic with $[Rh(bpy)_2]^+$ and $[Rh(phen)_2]^+$.

The nature of the species producing the g $\sim$ 4 signal is not obvious. The intensity of the line is concentration dependent which suggests the presence of an S = 1 dipole-dipole dimer. The position and shape of the line suggest that it results from forbidden $\Delta m = 2$ transitions associated with an S = 1 species. Cyclic voltammetry did not indicate the presence of a dimer but the reduction potential of a weak dipole-dipole dimer may not differ enough from that of the monomer to permit voltammetry to distinguish the species.

4. DISCUSSION

Correlation between electrochemistry and spectroscopy. The observation of stable delocalized species resulting from the addition of one electron to $[IrL_2Cl_2]^{1+}$ and $[RuL_3]^{2+}$ (26) is in accordance with predictions of the redox orbital character based on luminescence measurements; the failure to observe a stable species from a one electron reduction of $[RhL_2Cl_2]^{1+}$ is also

TABLE VII

Reduction Potentials* of bpy Complexes

Nominal e⁻ transfer	$[\text{Rh(bpy)}_3]^{3+}$	$[\text{Rh(bpy)}_2\text{Cl}_2]^{1+}$	$[\text{Ir(bpy)}_2\text{Cl}_2]^{1+}$	$[\text{Ru(bpy)}_3]^{2+a}$
$d^6 \to d^7$	-0.86	(-1.01)	-1.24	-1.33
$d^7 \to d^8$	-1.00	(-1.01)	-1.45	-1.52
$d^8 \to d^9$	-1.49	-1.47	-2.03	-1.76
$d^9 \to d^{10}$	-1.71	-1.68	NO	-2.40

*E_p in V vs aq. SCE; v = 31.6 V/sec

ᵃfrom (26).

TABLE VIII

Reduction Potentials* of phen Complexes

Nominal e^- transfer	$[Rh(phen)_3]^{3+}$	$[Rh(phen)_2Cl_2]^{1+}$	$[Ir(phen)_2Cl_2]^{1+}$	$[Ru(phen)_3]^{2+a}$
$d^6 \rightarrow d^7$	-0.80	(-0.95)	-1.32	-1.41
$d^7 \rightarrow d^8$	-1.06	(-0.95)	-1.53	-1.54
$d^8 \rightarrow d^9$	-1.54	-1.56	-2.10	-1.84
$d^9 \rightarrow d^{10}$	-1.69	-1.69	NO	-2.24

*E_p in V vs aq. SCE; v = 31.6 V/sec

[a] from (26).

consistent with our spectroscopic observations. The agreement between the two techniques breaks down for the $[RhL_3]^{3+}$ case. Indeed the rapid chemical reaction following the electron transfer
suggests production of the labile Rh(III) species, i.e. the redox
orbital is a d orbital.

The discrepancy could have its origin in the breakdown of the
orbital model (neglect of electron repulsion) or the breakdown of
the Koopman's theorem (27). Alternatively, the introduction of an
electron into a π^* orbital may in this case produce a chemically
unstable species.

Additional experimental data is required to clarify the origin
of this problem.

Electrochemistry and molecular orbitals. The total electron
transfer sequence for the Ru^{2+} and Ir^{+3} complexes and that for the
$[RhL_2]^n$ (where n = $\pm$ 1,0) show behavior that appears to justify the
application of Koopman's Theorem to the redox orbitals here.

The one-electron orbital energies for each metal ion system
appear to be independent of the number of electrons in the system.
Ultimately comparisons of the character of the molecular orbitals
for the three metal ion systems may be possible if experimental
conditions (sweep rates) for all three metal ion systems can be
standardized to permit meaningful comparisons of $E_{1/2}$'s and the
differences ($\Delta E_{1/2}$) in these values. Further, the $[Ir(L_2)Cl_2]^+$
species both show a peak at $\sim$ - 2V corresponding to a nominal
$d^8 \rightarrow d^9$ (Tables VII and VIII). The magnitude of the $E_{1/2}$ for this
process is not consistent with that observed for the $d^6 \rightarrow d^7$ and
$d^7 \rightarrow d^8$ process nor with the progression noted for the other complexes. The rationale of this anomaly will require verification
of these results and, likely, the voltammetry for the difficult to
synthesize $[IrL_3]^{3+}$ species.

5.   ACKNOWLEDGMENTS

This research was sponsored by the National Council of Science
and Technology of Romania, the National Science Foundation
(GF-40894) and North Carolina State University. Several figures
were reprinted with permission from the Journal of Physical
Chemistry.

6. REFERENCES

(1)   F. S. Hall and W. L. Reynolds, Inorg. Chem., 5, 931 (1966).

(2)   N. Tanaka and Y. Sato, Inorg. Nucl. Chem. Letts., 2, 359 (1966).

(3)   N. Tanaka and Y. Sato, Electrochim. Acta, 13, 335 (1968).

(4)   N. Tanaka and Y. Sato, Bull. Chem. Soc. (Japan), 41, 2059 (1968).

(5)   N. Tanaka and Y. Sato, Bull. Chem. Soc. (Japan), 41, 2064 (1968).

(6)   N. Tanaka, T. Ogata and S. Niizumi, Bull. Chem. Soc. (Japan), 46, 3299 (1973).

(7)   S. Musumeci, E. Rizzarelli, I. Fragala, S. Sammartano and R. P. Bonomo, Inorg. Chim. Acta, 7, 660 (1973).

(8)   S. Musumeci, R. Rizzarelli, S. Sammartano and R. P. Bonomo, J. Inorg. Nucl. Chem., 36, 853 (1974).

(9)   T. Saji and S. Aoyagui, Chem. Lett., 203 (1974).

(10) T. Saji and S. Aoyagui, J. Electroanal. Chem., 58, 401 (1975).

(11) T. Saji and S. Aoyagui, J. Electroanal. Chem., 60, 1 (1975).

(12) T. Saji, T. Yamada and S. Aoyagui, J. Electroanal. Chem., 61, 147 (1975).

(13) T. Saji and S. Aoyagui, J. Electroanal. Chem., 63, 31 (1975).

(14) J. E. Hillis and M. K. DeArmond, J. Chem. Phys., 54, 2247 (1971).

(15) J. N. Demas and G. A. Crosby, J. Amer. Chem. Soc., 93, 2841 (1971).

(16) G. Kew, K. DeArmond and K. Hanck, J. Phys. Chem., 78, 727 (1974).

(17) B. Martin, W. R. McWhinnie and G. M. Waind, J. Inorg. Nucl. Chem., 23, 207 (1961).

(18) G. Kew, K. Hanck and K. DeArmond, J. Phys. Chem., 79, 1828 (1975).

(19)   H. Caldararu, M. K. DcArmond, K. W. Hanck and V. E. Sahini,
       J. Amer. Chem. Soc., 98, 4455 (1976).

(20)   W. Halper and M. K. DeArmond, J. Lumin., 5, 225 (1972).

(21)   C. R. Wilson and H. Taube, Inorg. Chem., 14, 405 (1975).

(22)   R. D. Gillard, Coord. Chem. Reviews, 16, 67 (1975).

(23)   B. R. McGarvey, Trans. Met. Chem., 3, 181 (1966).

(24)   A. J. Carty and P. C. Chieh, Chem. Commun., 356 (1972).

(25)   M. Hinamota, S. Ooi and H. Kuraya, Chem. Commun., 356 (1972).

(26)   N. E. Takel-Takvoryan, R. E. Hemingway and A. J. Bard,
       J. Amer. Chem. Soc., 95, 6582 (1973).

(27)   I. Hanazaki and S. Nagakura, Bull. Chem. Soc. (Japan), 44,
       2312 (1971).

(28)   A. A. Vlcek, Electrochim. Acta, 13, 1063 (1968).

(29)   A. A. Vlcek, Proc. Int. Coord. Chem., 14, 220 (1972).

ACID-BASE DEPENDENT REDOX CHEMISTRY IN MOLTEN CHLOROALUMINATES

Gleb Mamantov* and Robert A. Osteryoung†

*Department of Chemistry, The University of Tennessee,
Knoxville, Tennessee   37916
†Department of Chemistry, Colorado State University,
Fort Collins, Colorado   80523

1.   INTRODUCTION

The Hall process for the production of aluminum was discovered 91 years ago; hence, the use of molten salts for industrial applications is certainly not new. However, only within the last 25 years has the chemistry of melts attracted substantial interest. Applications, such as the molten salt nuclear reactor, extraction of metals from ores, electrorefining, molten salt batteries, catalysis and uses in the glass industry, have stimulated (and helped to support) the fundamental studies of the properties of melts as well as of solute chemistry in melts. Molten halides are probably the melts used most frequently. In the past, the choice of melts for use as solvents was frequently based on such considerations as availability, ease of purification and the lowest liquidus temperature attainable, as well as the width of the electrochemical span and the spectroscopic transparency of the melt.

In recent years, we have become interested in melts whose composition can be adjusted to change the acid-base properties of the solvent system which, in turn, determine the redox and the coordination chemistry in these media. Aluminum halide-containing melts represent a prime example of systems in which the change in composition can greatly affect the acidity and the chemistry of solutes. The low-melting chloride melts have been studied most intensively. By varying the $AlCl_3/MCl$ (M^+ - alkali metal cation) molar ratio from somewhat less than one (the usual limitation is the solubility of MCl in $MAlCl_4$ at the typical working temperatures of 150-200°C) to greater than two, the modified Lewis acidity

223

(acid is defined as the chloride ion acceptor; the acidity is usually expressed in terms of $pCl^- \equiv - \log [Cl^-]$) and the concentrations of the species present ($AlCl_4^-$, $Al_2Cl_7^-$, Cl^-) are varied widely. Raman studies (1-3) have been used to define the species present and potentiometric measurements (4-7) have been employed to study quantitatively the equilibria involved in chloroaluminates. For $AlCl_3$ -NaCl molten mixtures, the predominant equilibrium near the equimolar point is $2AlCl_4^- = Al_2Cl_7^- + Cl^-$ (4-7). The mole fraction equilibrium constant for this equilibrium at 175°C is 1.1×10^{-7}. Thus, $AlCl_4^-$ is the predominant anion in equimolar $AlCl_3$-NaCl melt. In more acidic ($AlCl_3$-rich) melts the $Al_2Cl_7^-$ ion becomes more abundant as evidenced by both potentiometric and Raman studies. The most basic sodium chloroaluminate system corresponds to a NaCl-saturated melt in which pCl^- is constant in the presence of solid NaCl and at 175°C is equal to 1.1 (8). Variation of the equilibrium constant with temperature and with different counter ions has also been studied (9).

2. A HISTORICAL ACCOUNT AND A COMPARISON WITH SUPERACID SYSTEMS

The interest in the chemistry in molten chloroaluminates arose from studies of interactions between metals and their salts in the molten state (10,11). More specifically, the observations that certain halides, such as $ZnCl_2$ and NaCl, change the solubility of bismuth and cadmium in their halides (12,13) were reinterpreted (14) by assuming acid-base interactions between the added salt and the two oxidation states present, e.g. Cd^{2+} and Cd_2^{2+}. According to Corbett and McMullan (14) "added base, i.e. halide ion, would be expected to reduce the amount of subhalide formed through stabilization of the more acidic, higher oxidation state. Conversely, addition of an acid capable of complexing halide ion would increase the amount of subhalide formed." Using the strongly acidic $AlCl_3$, Corbett and McMullan prepared solid $BiAlCl_4$ (from the reaction $BiCl_3 + 2Bi + 3AlCl_3 = 3BiAlCl_4$) and $GaAlCl_4$ (using the reaction $GaGaCl_4 + 2Ga + 4 AlCl_3 = 4 GaAlCl_4$) and increased the conversion of $Cd + CdCl_2$ to Cd_2Cl_2 from 17.6% at 740° to 71.5% at 330° using a mole ratio of $2AlCl_3/CdCl_2$. This work was extended by the characterization of Ga_2Cl_4 and its analogy with $GaAlCl_4$ (15) and by the isolation of $Cd_2(AlCl_4)_2$ (16).

The in situ characterization of subvalent species in chloroaluminate and chlorozincate melts was greatly advanced by the skillful application of UV-visible spectroscopy by Bjerrum, Boston and Smith (17) to dilute solutions of the products of the reaction between bismuth and $BiCl_3$. These workers studied quantitatively the equilibria

$$6\ Bi^+_{(soln)} = Bi_5^{3+}{}_{(soln)} + Bi^{3+}_{(soln)}$$

$$2\ Bi_{(1)} + Bi^{3+}_{(soln)} = 3\ Bi^+_{(soln)}$$

$$4\ Bi_{(1)} + Bi^{3+}_{(soln)} = Bi_5^{3+}{}_{(soln)}$$

In addition to Bi^+ and Bi_5^{3+}, Bjerrum and Smith [18] characterized Bi_8^{2+} in the presence of excess bismuth metal.

Munday and Corbett [19] carried out EMF measurements on the cell Ag/Ag^+ in $NaAlCl_4/glass/M^{2+}$, $M_m^{(2-n)+}$ in $NaAlCl_4/Ta$ to study the lower oxidation states of cadmium, lead and tin in a slightly acidic $AlCl_3$-$NaCl$ melt. Formation of Pb^+ (or $Pb^+ \cdot n\ Pb^{2+}$), Cd_2^{2+} and Sn_2^+ (or Sn_3^{2+}) was indicated.

In 1968 Tremillon and Letisse published two papers [4,20] on the acid-base equilibria in $AlCl_3$ - $NaCl$ melts. They found that the aluminum electrode functions as a pCl^- indicator according to the reaction $AlCl_4^- + 3e^- = Al + 4\ Cl^-$. This work was later extended by Torsi and Mamantov [5,9] Osteryoung and coworkers [6], and Fannin, King and Seegmiller [7].

Mamantov and coworkers [21] used voltammetric, UV-visible and Raman spectral measurements to characterize the subvalent mercury species Hg_3^{2+}. The structure of solid $Hg_3(AlCl_4)_2$ was determined by means of single crystal X-ray diffraction measurements [22].

Extensive studies on the polyatomic ions of tellurium, selenium and sulfur in acidic chloroaluminates have been carried out by Bjerrum and coworkers [23-27] (see the chapter by Bjerrum). Solids containing these ions, such as $Te_4(AlCl_4)_2$ and $Se_8(AlCl_4)_2$, have been prepared and characterized by Corbett and coworkers [28,29]. Polyatomic cations of iodine with $AlCl_4^-$ counter ions have been prepared as solids by Merryman et al. [30,31].

In addition to presenting a fertile medium to inorganic chemists, molten chloroaluminates (and other acidic melts) are becoming of considerable interest to organic chemists, particularly those interested in stabilizing organic radical cations. Extensive contributions have been made by Osteryoung's group. This area has been recently reviewed by Jones and Osteryoung [32]; recent developments in organic electrochemistry in chloroaluminate melts are discussed below.

A review on the molten salt chemistry of the haloaluminates was prepared by Boston [33] in 1971.

A reasonably complete summary of "unusual" inorganic species characterized in molten chloroaluminates, as well as of solid compounds containing the $AlCl_4^-$ or $Al_2Cl_7^-$ ions, is given in Table I. This Table also includes a compilation of similar species found in superacid systems, such as $HSO_3F\text{-}SbF_5$. Gillespie (34) and Gillespie and Passmore (35) have reviewed the chemistry in superacid media and the homopolyatomic cations produced in these media. It is clear that the chemistry, both inorganic and organic, in molten chloroaluminates bears considerable similarity to that in superacid systems. Differences exist, however, as noted by Corbett in a recent review (36).

3.   RECENT DEVELOPMENTS IN THE CHEMISTRY OF INORGANIC SOLUTES

The discussion below is not intended to be comprehensive; it summarizes primarily the research that has been performed by our groups in the last few years.

Summaries of various systems studied in these melts are given in the book by Plambeck (50; literature appears to have been covered through 1975) and in the chapter by Fung and Mamantov (60; literature covered through mid-1972).

3.1 Group VIA

The behavior of oxide ions in these melts has long been of interest, but until recently few studies have been carried out. Even the reactivity of oxygen with the chloroaluminate melts--at least at 200°C or less--seems to be the subject of confusion. Oxide, for example, is electrochemically inert (although see below). However, the anions directly below O^{2-} in the periodic table, the chalcogenides, are electrochemically active.

The electrochemistry of sulfur has been studied by the groups of Osteryoung and Mamantov (61-63). Elemental sulfur is quite soluble in the melt; it is quantitatively reduced to sulfide. Cyclic voltammetry of the S/S^{2-} system indicates considerable electrochemical irreversibility. Sulfur can be oxidized at potentials close to those producing Cl_2 evolution; coulometric and voltammetric measurements in a basic melt at 175°C indicated the product to be S_2Cl_2 (61,63). At higher temperatures (∿230°C) S(II) is apparently the predominant oxidation product (63). No evidence for higher oxidation states was obtained in basic melts. In acidic melts, such as $AlCl_3$ - $NaCl$ (63 mole % $AlCl_3$), the electrooxidation of sulfur apparently involves three steps (64). Both normal pulse voltammetric and controlled-potential coulometric measurements indicated that the final oxidation product of sulfur

Table 1

"Unusual" Species Observed in Chloroaluminates or in Superacid Systems*

Group	Species or Compounds	Medium	References
IIB	Cd_2^{2+} , $Cd_2(AlCl_4)_2$	$NaAlCl_4$	19,36
IIB	Hg_3^{2+} , $Hg_3(AlCl_4)_2$	Acidic chloroaluminates ($AlCl_3/NaCl>1$)	21,22,36
IIB	$Hg_3(AsF_6)_2$, $Hg_3(Sb_2F_{11})_2$, $Hg_4(AsF_6)_2$, $Hg_{2.85}(AsF_6)_2$	SO_2 + AsF_5 (or SbF_5)	35-39
IVA	Sn_2^+ , Sn_3^{3+}	$NaAlCl_4$	19,36
IVA	Pb^+ , Pb_2^{3+}	$NaAlCl_4$	19,36
VA	Subvalent Sb Sb_n^{n+} , $SbAsF_6$	Sb + $SbCl_3$ + $AlCl_3$	36
		SO_2 + AsF_5	35,40,41
VA	$SbSO_3F$	HSO_3F	
VA	Bi^+ , Bi_5^{3+} , Bi_8^{2+}	Acidic chloroaluminates	17,18,36

*(Some compounds prepared using constituents of these systems are included).

Table 1 (continued)

Group	Species of Compounds	Medium	References
VIA	O_2AsF_6, O_2SbF_6, $O_2Sb_2F_{11}$	$O_2 + F_2 + AsF_5$ $O_2F_2 + SbF_5$	35,42-44
	S_{16}^{2+}, S_8^{2+}, S_4^{2+}, $S(II)$	Acidic chloroaluminates	45
	S_{16}^{2+}, S_8^{2+}, S_4^{2+}, S_5^+ $S_{16}(AsF_6)_2$, $S_8(Sb_2F_{11})_2$	HSO_3F, $HSO_3F + SbF_5$	35,46-48
	Se_{16}^{2+}, Se_{12}^{2+}, Se_8^{2+}, Se_4^{2+} $Se_8(AlCl_4)_2$	Acidic chloroaluminates	26,36
	Se_8^{2+}, Se_4^{2+} $Se_4(AsF_6)_2$, $Se_8(Sb_2F_{11})_2$	HSO_3F SO_2 or $HF + AsF_5$ or SbF_5	35,49,50
	Te_8^{2+}, Te_6^{2+}, Te_4^{2+}, Te_2^{2+} (?), $Te(II)$ (?), $Te_4(AlCl_4)_2$,	Acidic chloroaluminates	23,28,36

Table 1 (continued)

Group	Species of Compounds	Medium	References
	$Te_4(Al_2Cl_7)_2$		
	Te_4^{2+}, Te_2^{2+}	HSO_3F	35,51,52
	$Te_4(AsF_6)_2$	$SO_2 + AsF_5$	
VIIA	Cl_3AsF_6	$Cl_2 + ClF + AsF_5$ (-70°)	35,53
VIIA	Br_3^+, Br_2^+, Br_3AsF_6, $Br_2Sb_3F_{16}$	$HSO_3F + SbF_5 = SO_3$	35,54-56
VIIA	I_5^+, I_3^+, I_2Cl^+, I_5AlCl_4, I_3AlCl_4, $I_2ClAlCl_4$	$ICl + I_2 + AlCl_3$	30,31,36
VIIA	I_5^+, I_3^+, I_2^+, $I_2Sb_2F_{11}$	HSO_3F, $HSC_3F + SbF_5$	35,57,58

in acidic chloroaluminates is S(IV). Spectrophotometric and potentiometric studies by Bjerrum and coworkers (45) as well as Raman
studies (65) support the stability of S(IV) in these melts; both
potentiometric and Raman studies indicated that S(IV) is present as
SCl_3^+. $SCl_3^+AlCl_4^-$ (66) was prepared and characterized; addition
of SCl_3AlCl_4 to acidic melts led to similar electrochemistry as
obtained with the final oxidation product of sulfur. The reduction
of SCl_3^+ to sulfur apparently involves the intermediate species
S_2^{2+} and S_8^{2+} (64).

The S/S^{2-} peak potentials in cyclic voltammetry vary considerably with melt acidity. As a result of efforts to determine
S^{2-} "in situ" in the melt by coulometric titration with electrochemically generated Cu(I), Paulsen and Osteryoung (61) found that
S^{2-} underwent considerable acid-base chemistry, similar to that
occurring in aqueous systems. Metal sulfides - Cu_2S, Ag_2S, FeS -
were insoluble in basic melts, but readily soluble in acidic melts.
From coulometric experiments, in which the ratio of S/S^{2-} was
varied at fixed acidity, coupled to potentiometric-Nernst plot
measurements of the S/S^{2-} equilibrium potential, it was concluded
that the $\frac{RT}{4F}$ slope obtained indicated that the reaction

$$S_2 + 4e^- = 2 S^{2-}$$

was potential-determining at an inert electrode. By measuring the
S_2/S^{2-} potential at fixed concentrations of sulfur and sulfide,
but varying acidity, Paulsen and Osteryoung (61) concluded that

$$S^{2-} + AlCl_4^- = AlSCl + 3Cl^-$$

was the dominant acid-base equilibrium in acidic melts.

As the melt was made progressively more basic, this equilibrium shifted to the left. The initial experiments by Paulsen and
Osteryoung indicated that S^{2-} might exist as $AlSCl_2^-$ (a dibase) in
basic melts, but the data were too imprecise to permit unambiguous
interpretation in terms of other than the tribasic AlSCl. Titration of sulfide with $AlCl_3$, using melts of varying initial pCl^-
values, indicated that S^{2-} behaved as a tribase ($\bar{n}=3$, where $\bar{n}=$
$AlCl_3/S^{2-}$) at pCl^- values >2.8 but showed a decrease in basicity
as the melt became more basic. An $\bar{n}$ value of 2.5 (equal amounts
of AlSCl and $AlSCl_2^-$) was achieved at a pCl^- value of ~2.5 (67).

Potentiometric measurements on the Se/Se^{2-} couple indicated,
similarly to the S/S^{2-} system, that the potential-determining
reaction could be expressed as

$$Se_2 + 4e^- = 2Se^{2-}$$

Nernst plots for the Se_2/Se^{2-} system at fixed concentrations of Se and Se^{2-} but varying pCl^- values resulted in two straight line segments with slopes $3RT/2F$ for $pCl^- > 3.0$ and RT/F for $pCl^- < 3.0$. This was interpreted as showing that AlSeCl is the dominant species at $pCl^- > 3.0$, since the electrode reaction could be taken as

$$Se_2 + 2AlCl_4^- + 4e^- = 2 \ AlSeCl + 6 \ Cl^-$$

for which the Nernst expression is

$$E = E^\circ - \frac{RT}{4F} \ln \frac{[AlSeCl]^2[Cl^-]^6}{[Se_2] [AlCl_4^-]^2}$$

If it is assumed that the concentration of all species except Cl^- is fixed then the equation simplifies to

$$E = E^{\circ\prime} - \frac{3RT}{2F} \ln [Cl^-] = E^{\circ\prime} + \frac{3RT}{2F} \cdot 2.3 \ pCl^- = E^{\circ\prime} + .133 \ pCl^-$$

at $175°C$

Experimentally, the slope was found to be 130 mV. Similarly, the RT/F slope indicated that $AlSeCl_2^-$ is the dominant species in basic melts. The intersection of the two lines at $pCl^- \sim 3.0$ corresponds to $\bar{n} = 2.5$ (equal amounts of AlSeCl and $AlSeCl_2^-$) and indicates that S^{2-} is a stronger base than Se^{2-}. The calculation of the equilibrium constant for $AlSeCl_2^- = AlSeCl + Cl^-$ is also possible. The titration experiments showed that Te^{2-} behaves similarly to Se^{2-} and S^{2-}.

The oxidation of selenium in acidic melts occurs at a potential $\sim0.1V$ less positive than that of sulfur (62).

As noted earlier, the behavior of oxide in these melts has been of considerable interest. Lettise and Tremillon (20) reported that OH^- (O^{2-}) behaved as a tribase in the chloroaluminate melts, forming AlOCl (or possibly $AlOCl \cdot AlCl_4^-$ or $Al_2OCl_5^-$). Recently, however, Tremillon, Bermond and Molina (68) reported that O^{2-} behaved as a dibase, in contrast to their earlier work (20). The work of Osteryoung and coworkers (67) had indicated that S^{2-}, Se^{2-} and Te^{2-}, as well as O^{2-}, behaved as either di- or tri-bases, depending on melt acidity. Further, all behaved as tribases in melts when the base was added to an intially neutral melt. In addition, it has been demonstrated that metal oxides and chalcogenides react with $AlCl_3$ in sealed tubes at temperatures around

300°C to yield AlOCl, AlSCl, AlSeCl and AlTeCl (69-71).

In addition to reporting that O^{2-} was dibasic, Tremillon, Bermond and Molina (68) reported on the first known determination of O^{2-} in the molten chloroaluminate system. This was based on the anodic depolarization of a Ni electrode by O^{2-} to form NiO. The voltammetric (chronoamperometric) waves obtained were pCl^- dependent and were claimed to be diffusion controlled in oxide. In basic melts one wave, stated to be due to the reaction

$$Ni + AlOCl_x + (4-x) Cl^- = NiO(s) + AlCl_4^- + 2e^-$$

was found; as the melt was made more acidic, two waves, diffusion limited in both O^{2-} and Cl^-, were found. Eventually, the reaction

$$Ni + AlOCl_x + (7-2x) AlCl_4^- = NiO(s) + (4-x) Al_2Cl_7^- + 2e^-$$

occurred again, yielding one O^{2-} diffusion-limited wave. Based on this reaction, the potential of a Ni/NiO electrode should be given by

$$E = E° + \frac{2.3RT}{2F} (4-x) pCl^- - \frac{2.3RT}{2F} \log [O^{2-}]$$

Experiments at fixed pCl^- values, in which E was plotted against $\log [O^{2-}]$, yielded a linear plot with slope very close to that predicted (48 mV at 210°C). At fixed $[O^{2-}]$, measurements of the Ni/NiO potential in pCl^- yielded two linear regions. The first, $pCl^- < 4.3$, had a 96 mV slope, which indicated x in the equation above was 2; i.e., $AlOCl_2^-$ exists as the dominant O^{2-} species. At $pCl^- > 4.3$, the slope was ~145 mV, indicating that x = 1, or that the dominant O^{2-} species was AlOCl.

For a number of reasons, Gilbert and Osteryoung (72) reinvestigated this topic. In initial work, immediately after publication of the recent paper by Tremillon et al., Robinson and Osteryoung (73) were unsuccessful in reproducing the Ni electrode voltammetric behavior reported by Tremillon et al. (68). Further, earlier work (20) had indicated O^{2-} to be tribasic, at least in the neutral melt, while Tremillon, Bermond and Molina (68) maintained that O^{2-} was a dibase to a pCl^- of ~4.3--well into the acid range. Gilbert and Osteryoung (72) found, in contrast to Tremillon et al., that it was $NiCl_2$, not NiO, that precipitated when an acid solution of Ni(II) was made basic by addition of NaCl. Cyclic voltammetry was used to follow the decrease in the amount of Ni(II) with pCl^-, and the solubility product of $NiCl_2$ was determined to be $10^{-12.8}$ (mole $kg^{-1})^3$ at 175°C. X-ray analysis of the precipitate verified that it was $NiCl_2$. The potential of a Ni/$NiCl_2$ electrode was monitored as a function of pCl^-. The slope was found to be 91 mV as predicted by the expression

$$E = E^{\circ}_{Ni/Ni(II)} + \frac{2.3RT}{2F} \log K_{sp} + \frac{2.3RT}{2F} 2 \, pCl^-$$

where the factor in front of the pCl^- term is 89 mV at 175°C. From the E° value for Ni/Ni(II), determined in acid mixtures to be 1.445 V vs. Al, the K_{sp} for $NiCl_2$ was calculated to be 10^{-13} (mole $kg^{-1})^3$, in agreement with the cyclic voltammetric data. Then, the standard potential of the $Ni/NiCl_2/Cl^-$ second class electrode is 0.87 V vs Al.

Also in contrast to the results of Tremillon et al. (68), Gilbert and Osteryoung (72) found that the potential of a Ni electrode (slightly anodized for stability) was <u>independent</u> of added O^{2-} if the pCl^-, measured at an Al electrode, was readjusted to its initial value after each O^{2-} ($BaCO_3$) addition.

Precise titration of O^{2-} added to melts at various pCl^- values by coulometric generation of Al(III) from an Al electrode indicated that O^{2-} was a tribase to a pCl^- of ∿2.7 and that $\bar{n}$ then decreased to 2.5 at a pCl^- of ∿1.9. Thus, O^{2-} was indicated to be a stronger base than S^{2-}, Se^{2-} or Te^{2-}. Equilibrium constants were obtained for the reactions:

$AlCl_4^- + O^{2-} = AlOCl + 3Cl^-$ $K_1 = (8.4 \pm 1.1) \; 10^{-6}$ (mole $kg^{-1})^2$

$AlCl_4^- + O^{2-} = AlOCl_2^- + 2 \, Cl^-$ $K_2 = (5.2 \pm 1) \; 10^{-4}$ mole kg^{-1}

$AlOCl_2^- = AlOCl + Cl^-$ $K = K_1/K_2 = 1.0 \pm 0.5 \times 10^{-2}$ mole kg^{-1}

The results stated above force the conclusion that the behavior ascribed by Tremillon et al. (68) to O^{2-} depolarization of Ni was in reality due to Ni depolarization by chloride. Thus, no viable method for a determination of oxide in these melts yet exists.

3.2 Refractory Metals

The electrochemistry of titanium, zirconium, hafnium, niobium, tantalum, molybdenum and tungsten in chloroaluminate melts of different compositions has been studied by various electrochemical techniques. A good example of the importance of the acid-base properties of the chloroaluminates involves the Ti(II) - Ti(III) - Ti(IV) redox system (74). The $E_{1/2}$ values for both redox steps

shift to more cathodic potentials with even relatively small decreases in pCl⁻. Ti(II), while perfectly stable in the very
acidic AlCl$_3$ - NaCl (63 mole % AlCl$_3$) melt, precipitates from the
52 mole % AlCl$_3$ melt.

The reduction pathways of the highest oxidation states of
zirconium, niobium, tantalum and tungsten are dependent on melt
composition (or Lewis acidity) and temperature. Significant
differences in the redox chemistry of these elements were observed,
as outlined below.

The electrochemical reduction of Zr(IV) (75), added as either
ZrCl$_4$ or Na$_2$ZrCl$_6$, in molten AlCl$_3$ - NaCl(51-52 mole % AlCl$_3$)
resulted in the formation of insoluble ZrCl$_3$, which exhibited a
strong Raman line which is probably due to a Zr-Zr bond. On the
other hand, the reduction of Zr(IV) in much more acidic melts,
such as AlCl$_3$ - NaCl (60 mole % AlCl$_3$), resulted in the formation
of Zr(II) at high temperatures (> 250°C) and of soluble Zr(III) at
low temperatures (< 140°) which apparently disproportionates to
Zr(IV) and Zr(II). In the NaCl-saturated melt (~ 49.8 mole % AlCl$_3$
at 175°C) ZrCl$_4$ and Na$_2$ZrCl$_6$ were insoluble. It is interesting
that the reduction of Zr(IV) in 51-52 mole % AlCl$_3$ involves nucleation overpotential, as indicated by several electrochemical
criteria. This effect had not been previously observed in molten
halides; in fact, Hills et al. (76) had noted that it was unlikely
that nucleation effects would be observed in molten halides, where
predeposition reactions to form a monolayer frequently occur (77).

As expected from the comparison of the electrochemical behavior
of Ti(IV) (74) and Zr(IV) in molten chloroaluminates, the electrochemical reduction of Hf(IV) in slightly acidic (∿ 51 mole % AlCl$_3$)
melts occurred at very cathodic potentials, after the Al$_2$Cl$_7^-$
reduction (see below) but before the cathodic limit (reduction of
AlCl$_4^-$) (78). The reduction product, characterized by x-ray powder
diffraction, was the alloy HfAl$_3$. In the basic (NaCl-saturated)
melt NfCl$_4$ was insoluble; in very acidic melts Al$_2$Cl$_7^-$ reduction
became the cathodic limit of the melt. The difference in electrochemical reduction potentials of Zr(IV) and Hf(IV) in 51-52 mole %
AlCl$_3$ is sufficient to separate these two elements, particularly
since ZrCl$_3$ is insoluble. The separation of hafnium and zirconium,
made difficult by their very similar chemistry, has attracted
considerable attention over the years (78). We have found (79)
that the precipitate of ZrCl$_3$ obtained by controlled-potential deposition was free of hafnium, as determined by x-ray fluorescence.
Further studies of this separation are currently in progress (80).

The electrochemistry of Nb(V) is quite involved (81). For
example, in the NaCl-saturated melt the mechanism believed to be
present is

$$Nb^{5+} + e^- = Nb^{4+} \qquad 2Nb^{4+} \overset{k_f}{\underset{k_b}{=}} Nb_2^{8+} \qquad Nb_2^{8+} + e^- = Nb_2^{7+}$$

$$Nb_2^{7+} + e^- = Nb_2^{6+} \qquad 3Nb_2^{6+} + 2e^- = 2Nb_3^{8+}$$

Recent studies (82) showed that the niobium cluster produced was insoluble. The reduction mechanism was different in acidic melts as well as in basic melts containing added oxide ions (83). However, additions of oxide ions to very acidic melts did not result in changes in the voltammetric behavior of Nb(V) (83).

The chemistry of tantalum in chloroaluminate melts is being pursued currently (84). Ta(V) is considerably more difficult to reduce than Nb(V) in these melts. The electrochemical reduction proceeds through Ta(IV) (which can be observed only at low temperatures) and the unstable Ta(III) to the cluster $Ta_6Cl_{12}^{2+}$. The cluster is not very soluble in the 63 mole % $AlCl_3$ melt and quite insoluble in the NaCl-saturated melt.

The redox chemistry of molybdenum (85) and tungsten (86) in these media is very complex. The electrochemistry of molybdenum has been studied in the basic melt system (85). The Mo(III)-Mo(V) couple gave a reversible Nernst slope; the potential change as a function of melt acidity indicated that the reaction is

$$Mo(III) + 2 Cl^- = Mo(V) + 2e^-$$

The exact nature of the Mo(III) species, however, was not determined (85). Mo(V) was reduced to Mo(III), with indications that Mo(III) oxidation involves a disproportionation while the reduction of Mo(V), as deduced from rotating disc experiments, suggests a follow-up chemical reaction involving chloride. Considerable surface chemistry is involved. All Mo(VI) species oxidized the melt to yield Cl_2 and Mo(V). It appears, based on spectrophotometric evidence, that the oxygenated Mo(VI) species is reduced to a Mo(V) species which does not contain oxygen. This is a consequence of the strong tendency of the melt to react with O^{2-} to form $AlOCl$ or $AlOCl_2^-$ (see above). For the tungsten system, formation of cluster species, such as $W_6Cl_8^{4+}$, by the stepwise electrochemical reduction of W(VI) was indicated (86).

The reduction potential of the highest oxidation state becomes more negative in going down each periodic table group and more positive in going from Group IVB to Group VIB. No definite evidence for the deposition of any metals studied from chloroaluminate melts has been obtained, although approximate thermodynamic calculations (87) predict that deposition of most refractory metals

from these melts should be expected. Since the reduction of
several metals involves formation of metal clusters (some of which
are soluble in these melts), it appears that a significant reaction
overpotential is involved in the reduction of the cluster to the
metal (this overpotential may be required to rearrange the crystal
lattice or to expel bridging chloride ions from the cluster).
Alternatively, the formation of insoluble lower valent halides may
impede the reduction to the metal. It is interesting to contrast
the electroreduction of refractory metal ions in molten chloroaluminates with that in molten fluorides (88). In fluorides
clusters are not stable; this fact coupled with the large cathodic
"window" in molten alkali fluorides has resulted in the development of plating of coherent deposits of refractory metals from
fluoride melts (89).

3.3 Miscellaneous

     The electrochemical behavior of aluminum in chloroaluminate
melts has been studied primarily with aluminum electrodes (20,90-
93). In $AlCl_3$-rich melts the oxidation of aluminum results in the
formation of a passivating Al_2Cl_6 layer at the electrode surface
(91,92). Although several Russian workers had reported that univalent aluminum ions were formed upon anodization of the aluminum
electrode in the $AlCl_3$ - NaCl systems, recent work under carefully
controlled conditions has failed to substantiate these findings
(93). Rolland and Mamantov have examined the electrochemical
reduction of $Al_2Cl_7^-$ ions at foreign substrates in $AlCl_3$-NaCl
melts containing 50.1-51 mole % $AlCl_3$ (94). It was found that the
reduction of $Al_2Cl_7^-$ ions occurs at potentials ∿0.4V positive with
respect to the melt limit (reduction of $AlCl_4^-$); the reduction
involves a nucleation process which is particularly pronounced at
glassy carbon electrodes. At platinum electrodes alloy formation
is also involved. The calculated diffusion coefficients for
$Al_2Cl_7^-$ at 175°C ranged from 3.2 to 9.6×10^{-6} cm^2/sec depending on
the method (chronopotentiometry, chronoamperometry and cyclic
voltammetry) and the electrode (glassy carbon, platinum and
tungsten) used. Similar scatter was observed by Hills and coworkers (76) for the reduction of Ag^+ in molten nitrates; the
observed discrepancies were also attributed to the nucleation
effect.

     Cu(I) was found, on the basis of potentiometric measurements,
to form a complex best represented as $CuCl_2^-$ (95). Nernst plots
for the Cu-Cu(I) couple extrapolated to different E° values,
depending on melt acidity. Investigation of this variation in a
region very close to 1:1 $AlCl_3$:NaCl permitted a calculation of the
equilibrium constant for the reaction

$$Cu^I + 2 Cl^- = CuCl_2^- \qquad\qquad K_{Cu}$$

K_{Cu} was found to be $6.5 \times 10^4 \ M^{-1}$.

The strong acid character of the sodium tetrachloroaluminate system was demonstrated by Gale and Osteryoung (96), who showed that NO_3^- oxidized the melt to give Cl_2 and $NOCl$, and that NO_2^- reacted to give high yields of $NOCl$, presumably as a result of the reaction

$$NO_2^- + AlCl_4^- = NO^+ + AlOCl + 3Cl^-$$

A UV spectrum of the melt containing $NOCl$ had a broad absorption band with a λ_{max} of 255 nm, which is similar to that for the nitrosyl cation in aqueous systems (260 nm). Electrochemical studies of NO^+ indicated a reversible reduction to NO, with an E° of ~ 1.86 V vs Al (96).

Iodide is oxidized in two steps to iodine and $I(I)$ in both basic and acidic melts (97). The experimental $E_{1/2}$ values indicate that both I_2 and $I(I)$ are more strongly complexed in the NaCl saturated than in the acidic melts. The exact nature of $I(I)$ could not be determined from the results obtained.

4.   ORGANIC ELECTROCHEMISTRY

The behavior of organic compounds in chloroaluminate melts has been reasonably extensively studied, as perusal of a recent review article will indicate (32). However, organic electrochemistry in these solvents has not been extensively investigated, although in the past few years several papers have appeared. As noted earlier, marked changes in acidity take place in the region near the $NaAlCl_4$ composition and it has been found that certain organic compounds are very sensitive, when examined by a variety of electrochemical techniques, to changes in melt acidity (pCl^-).

Organic amines were chosen for investigation by Osteryoung's group because they were known to be electroactive in room temperature non-aqueous solvents, and also because it was assumed that the amines would interact with these strongly acidic solvent systems.

One of the first compounds to be studied was dimethylaniline (DMA) (98,99). Cyclic voltammetry on DMA on the first cycle resulted in one oxidation wave (if the scan was initiated from the equilibrium potential). Scan reversal yielded two reduction waves and a subsequent reversal yielded two oxidation waves. As the pCl^- was increased, corresponding to the system being made more acidic,

the first reduction peak grew at the expense of the second. The
behavior of the oxidation wave, as well as the dependence on scan
rate, suggested that the product of the oxidation underwent follow-
up chemistry. The first reduction peak found (see Fig. 1 of Ref.
99) is considered to be due to the reduction of the radical cation
of DMA; increasing acidity appears to stabilize the radical cation
relative to its subsequent coupling reaction to form N,N,N',N'-
tetramethylbenzidine (TMB). The addition of TMB to the melt
yielded a redox couple with cyclic voltammetric peaks at the potentials of the second (more negative) reduction wave and the first
(more negative) oxidation wave. This behavior indicated an e.c.e.
type mechanism.

While the detailed consideration of this process is beyond
the scope of this review, the net effect of increasing the acidity
is to stabilize the radical cation, perhaps by causing a decrease
in the rate of the second chemical step, C2.

The shift in the oxidation peak with acidity which was noted
in studying DMA indicated the presence of a preceding chemical
reaction as well; a detailed study of such a process was not
possible with DMA since the follow-up chemistry complicated the
situation. However, a study of N,N-dimethyl-p-toluidine, in which
the placement of a methyl group in the para position of DMA should
have prevented the coupling reaction, was carried out (99). As
the acidity of the melt was increased, the oxidation peak shifted
markedly anodically and became more drawn out, while the single
reduction peak (indicating that the radical cation was stabilized)
changed only slightly. This indicated rather conclusively that
acid-base dependent chemistry involving the solute did indeed

precede the electron transfer (oxidation) process.   Such behavior
may be described as

$$AlCl_4^- + B:AlCl_3 = Al_2Cl_7^- + B$$

$$B = B^+ \cdot + e^-$$

where B is the organic base.   The shift in the anodic oxidation
peak with increasing acidity indicated the rate of the reaction to
form B, the species undergoing oxidation at the electrode, was
diminished as the acidity was increased.   The ease of oxidation of
the two species, B and B:AlCl$_3$, where B:AlCl$_3$ is intended to
indicate the amine complexed by some acid moiety (not necessarily
AlCl$_3$), should be different, and under specific conditions it
should be possible to observe experimentally the oxidation of both
entities.   This possibility was demonstrated by studying the electrochemical oxidation of TMB (8).

      TMB in a NaCl saturated melt at a pCl of 1.1 showed one oxi-
dation wave at about 1.03 V against an Al reference in a NaCl
saturated melt and a single reduction wave.   The cyclic voltammogram was well-defined, although the peak separation was somewhat
greater than would be expected for a reversible two-electron
oxidation.   As the melt was made more acidic, a second wave appeared at about 2.0 V vs the Al reference.

      Normal pulse voltammograms for the oxidation of TMB at a pCl$^-$
of 3.4, obtained at two different pulse widths, showed two waves.
The ratio of the heights of the two waves varied with pulse width,
but the sum of the wave heights, as deduced from the time dependence of the total wave height, was diffusion controlled.   This
indicated an acid-base dependent chemical step preceding the first
wave.   The second wave, as discussed earlier for DMA, is due to
the oxidation of an entity which is harder to oxidize and presumably corresponds to the oxidation of the species indicated as
B:AlCl$_3$ in Equation 5 above.   The pulse voltammetric and cyclic
voltammetric behavior was indicative of the general scheme for a
preceding chemical reaction (a c.e. mechanism), which may be
written as:

$$Y \underset{k_b}{\overset{k_f}{\rightleftharpoons}} R \xrightarrow{-ne^-} 0$$

The pulse voltammetric data could be treated by the method of
Delahay (100) for the equations:

$$i_k/i_d = \pi^{1/2} \lambda \exp(\lambda^2) \, \mathrm{erfc}(\lambda)$$

where λ $(k_f Kt)^{1/2}$, i_k = kinetic current and i_d = diffusion current, with $K = [R]/[Y]$ and t equal to the pulse width.

Consistent values of $k_f'K_1$ could be obtained by studying the reaction at a series of pCl^- values and defining the present system, to reflect the acid-base dependent chemistry, as

$$Y + Cl^- \underset{k'_b}{\overset{k'_f}{\rightleftharpoons}} R + AlCl_4^-$$

$$R = 0 + 2e^-$$

with $K_1 = k'_f/k'_b$.

Since $K_1 = \dfrac{[R][AlCl_4^-]}{[Y][Cl^-]}$ and $k'_f = \dfrac{k_f}{[Cl^-]}$, then $k_f K = k'_f K_1 [Cl^-]^2/[AlCl_4^-]$

with $[AlCl_4^-]$ = 8.8 M in the $AlCl_3$:NaCl (50 mole % $AlCl_3$) melt.

Single potential step chronoamperometry was also employed to study the kinetics of the preceding reaction. The potential was stepped from a value negative of the first wave, where no reaction took place, to a potential beyond the first wave but not into the region where the second process took place. In basic (pCl^- < 1.5) region (the region where only one wave is observed in cyclic voltammetry), $it^{1/2}$ values were constant, indicating a diffusion controlled process. For pCl^- > 1.5, values of $it^{1/2}$ decreased as a function of time.

The $it^{1/2}$ data were also analyzed to obtain a value for $k_f'K_1$ for reaction (8). $k_f'K_1$ was found to be 4 x 10^6 liters mole^{-1} sec^{-1}, in agreement with the values obtained by analysis of the normal pulse voltammetric data. Reference (8) contains the details of these calculational procedures.

Since the electrochemical procedures yield only the product of an equilibrium constant, K_1, and a rate constant, k_f', spectro-photometry was employed to determine whether the equilibrium constant for Reaction 8 could be obtained. By comparison with studies on mono- and di-protonated TMB in acetonitrile, it was hoped that pCl^- variation would permit observation of free and complexed base (101). Saget and Plichon (101) found that the diprotonated TMB, in acetonitrile, had a λ_{max} at 250 nm, while the mono- and unprotonated forms had λ_{max} values of 320 and 310 nm. TMB in the $AlCl_3$ - NaCl system had a λ_{max} of 256 nm, independent of pCl^- over the range 1.1 to 4. Because the spectrum of biphenyl in the melt also had a λ_{max} of 253 nm, it was concluded that the band observed for TMB solution was due to a $\pi \rightarrow \pi^*$ transition in a TMB acid-base adduct, with the nonbonding electrons in both amino groups of TMB being shared with a Lewis acid, most likely $AlCl_3$, although bonding with other acid moieties in the melt cannot be ruled out.

The equilibria which best describe the situation are

$$TMB \ (AlCl_3)_2 + Cl^- \ \underset{\longleftarrow}{\overset{K_1}{\rightleftharpoons}} \ TMB \ (AlCl_3) + AlCl_4^-$$

$$TMB \ (AlCl_3) + Cl^- \ \underset{\longleftarrow}{\overset{K_2}{\rightleftharpoons}} \ TMB + AlCl_4^-$$

As the melt was made more acid, the rates of the dissociation reaction were decreased until, at $pCl^- > 1.5$, the oxidation of both the "free" TMB and the acid adduct, which is more difficult to oxidize, could be observed.

Analysis of the second oxidation wave indicated irreversible behavior, with slow electron transfer complicated by follow-up chemical reactions involving loss of $AlCl_3$ (or another acid moiety) from the oxidized adduct.

Tetrachloro-p-benzoquinone (chloranil) has also been studied (102). This material had been suggested by Mamantov and co-workers as a possible cathode for a molten salt battery system (103). Again, significant acid-base dependent electrochemistry was encountered. Chloranil is a strong oxidant (rest potential in $NaAlCl_4$ is 1.75V vs Al in a saturated melt). The melt turns red upon addition of chloranil. The extent of the oxidation is slight, however. Cyclic voltammetry at slow scan rates in a basic ($pCl^- = 1.1$) melt yielded a single reduction wave at $\sim$1.65 V vs Al. As the scan rate was increased, the first reduction wave shifted in the negative direction and a second wave appeared at $\sim$1.15 V. On scan reversal, however, only a single oxidation wave was observed at both fast and slow scan rates. These observations are consistent with the production of a semiquinone of limited stability, which undergoes a chemical reaction to give a product which can immediately be reduced. At fast scan rates the semiquinone itself is reduced because the sweep rate is faster than the chemical reaction.

Coulometry indicated a two-electron reduction to the dianion, which could be reoxidized. Normal pulse voltammetry also indicated two waves in basic melt at short pulse widths; as the melt was made more acidic only one wave was obtained. Thus, the semiquinone is rendered less stable in the acidic melt.

The overall mechanism appears best represented by:

$$\text{Cl} \quad \overset{+e^-}{\underset{-e^-}{\rightleftharpoons}} \quad \overset{+e^-}{\underset{-e^-}{\rightleftharpoons}}$$

$E_1 \qquad \qquad E_3$

$$\downarrow k \quad +AlCl_3$$

$$\overset{+e^-}{\underset{-e^-}{\rightleftharpoons}} \quad \xrightarrow{+AlCl_3}$$

E_2

As the melt is made more acid, the rate of the intervening chemical step increases until, at high acidity, no evidence for the semi-quinone can be found. These conclusions were verified by means of rotating disc studies, and a pseudo first-order rate constant of 55.5 cm sec^{-1} was found for the reaction of the semiquinone with the Lewis acid, assuming that E_2 was coincident with E_1 - i.e., that the equilibrium constant for the reaction . . .

$$+ \quad \overset{K_{eq}}{\rightleftharpoons} \quad + $$

was unity. The rotating disc electrode proved particularly useful for these studies since it minimized the high backgrounds typically encountered in cyclic voltammetry at glassy carbon electrodes in the melts.

The oxidation of tetrachlorohydroquinone was also studied, and it was concluded that the overall process is:

$$\bar{O} \rightarrow AlCl_3 \text{ (benzene ring) } O \rightarrow AlCl_3 \rightleftharpoons \text{(quinone)} + 2\ AlCl_3 + 2e^-$$

UV spectroscopy was interpreted to indicate that the $QCl_4{}^{2-}$ is dicomplexed with the acid moiety at all available pCl^- values.

The electrochemical oxidation of two sulfur heterocycles, tetrathioethylene and thianthrene, in acidic $AlCl_3$ - $NaCl$ mixtures was studied by Fung, Chambers and Mamantov (104). The radical cations and the dications of both compounds are readily prepared by electrochemical generation; the dications are considerably more stable in molten chloroaluminates at $150^{\circ}C$ than in dry acetronitrile at room temperature.

Very recently Demitras and Muetterties (105) have reported that $Ir_4(CO)_{12}$ catalyzes the hydrogenation of carbon monoxide in the acidic 2 $AlCl_3/NaCl$ melt to produce saturated hydrocarbons, such as propane. The previously mentioned possibility of forming metal clusters in the melts may mean that these systems have potential utility in organic catalytic reactions.

5.   ACKNOWLEDGMENTS

GM would like to acknowledge current support by the National Science Foundation and the Energy Research and Development Administration for studies in molten chloroaluminates. Prior support by the Atomic Energy Commission, Army Electronics Command and the Alcoa Foundation is also gratefully acknowledged. RAO would like to acknowledge currrent support from the Air Force Office of Scientific Research, the Office of Naval Research and a grant-in-aid from the Gould Foundation. Support in prior years from the American Chemical Society, Petroleum Research Fund, Army Research Office, Durham, the National Aeronautics and Space Administration and a grant-in-aid from the Alcoa Foundation are also acknowledged.

We are indebted and grateful to our coworkers for their contributions to this research as well as to G.P. Smith, Oak Ridge National Laboratory, and S.E. Springer, The University of Tennessee, for helpful comments pertaining to this manuscript.

6. REFERENCES

(1) G. Torsi, G. Mamantov, and G. M. Begun, Inorg. and Nucl. Chem. Letters, 6, 553 (1970).

(2) G. M. Begun, C. R. Boston, G. Torsi, and G. Mamantov, Inorg. Chem., 10, 886 (1971).

(3) E. Rytter, H. A. Øye, S. J. Cyvin, B. N. Cyvin and P. Klaeboe, J. Inorg. Nucl. Chem., 35, 1185 (1973).

(4) B. Tremillon and G. Letisse, J. Electroanal. Chem., 17, 371 (1968).

(5) G. Torsi and G. Mamantov, Inorg. Chem., 10, 1900 (1971).

(6) L. G. Boxall, H. L. Jones, and R. A. Osteryoung, J. Electrochem. Soc., 120, 223 (1973).

(7) A. A. Fannin, L. A. King, and D. W. Seegmiller, ibid., 119, 801 (1972).

(8) D. E. Bartak and R. A. Osteryoung, ibid., 122, 600 (1975).

(9) G. Torsi and G. Mamantov, Inorg. Chem., 11, 1439 (1972).

(10) J. D. Corbett, in "Fused Salts," B. R. Sundheim, ed., McGraw-Hill, New York, 1964, pp. 341-407.

(11) M. A. Bredig, in "Molten Salt Chemistry," M. Blander, ed., Interscience, New York, 1964, pp. 367-425.

(12) D. Cubicciotti, J. Amer. Chem. Soc., 74, 1198 (1952).

(13) G. Cleary and D. Cubicciotti, ibid., 74, 557 (1952).

(14) J. D. Corbett and R. K. McMullan, ibid., 78, 2906 (1956).

(15) R. K. McMullan and J. D. Corbett, ibid., 80, 4761 (1958).

(16) J. D. Corbett, W. J. Burkhard and L. F. Druding, ibid., 83, 76 (1961).

(17) N. J. Bjerrum, C. R. Boston and G. P. Smith, Inorg. Chem., 6 1162 (1967).

(18) N. J. Bjerrum and G. P. Smith, ibid., 6, 1968 (1967).

(19) T. C. F. Munday and J. D. Corbett, ibid., 5, 1263 (1966).

(20)   B. Tremillon and G. Letisse, J. Electroanal. Chem., 17, 387
       (1968).

(21)   G. Torsi, K. W. Fung, G. M. Begun and G. Mamantov, Inorg.
       Chem., 10, 2285 (1971).

(22)   R. D. Ellison, H. A. Levy and K. W. Fung, ibid., 11, 833
       (1972).

(23)   N. J. Bjerrum, ibid., 9, 1965 (1970); 10, 2578 (1971); 11,
       2648 (1972).

(24)   J. H. von Barner, N. J. Bjerrum and K. Kiens, ibid., 13,
       1708 (1974).

(25)   F. W. Poulsen, N. J. Bjerrum and O. F. Nielsen, ibid., 13,
       2693 (1974).

(26)   R. Fehrmann, N. J. Bjerrum and H. A. Andreasen, ibid., 14,
       2259 (1975).

(27)   R. Fehrmann, N. J. Bjerrum and H. A. Andreasen, ibid., 15,
       2187 (1976).

(28)   T. W. Couch, D. A. Lokken and J. D. Corbett, ibid., 11, 357
       (1972).

(29)   R. K. McMullan, D. J. Prince and J. D. Corbett, ibid., 10,
       1749 (1971).

(30)   D. J. Merryman, P. A. Edwards, J. D. Corbett, and R. E.
       McCarley, Chem. Commun., 779 (1972).

(31)   D. J. Merryman, P. A. Edwards, J. D. Corbett and R. E.
       McCarley, Inorg. Chem., 13, 1471 (1974).

(32)   H. L. Jones and R. A. Osteryoung, in "Advances in Molten
       Salt Chemistry," Vol. 3, J. Braunstein, G. Mamantov, and
       G. P. Smith, eds., Plenum Press, New York, 1975, pp. 121-176.

(33)   C. R. Boston, in "Advances in Molten Salt Chemistry," Vol. 1,
       J. Braunstein, G. Mamantov and G. P. Smith, eds., Plenum
       Press, New York, 1971, pp. 129-163.

(34)   R. J. Gillespie, Acc. Chem. Res., 1, 202 (1968).

(35)   R. J. Gillespie and J. Passmore, in "Advances in Inorganic
       Chemistry and Radiochemistry," Vol. 17, H. J. Emeleus and
       A. G. Sharpe, eds., Academic Press, 1975, pp. 49-89.

(36) J. D. Corbett, in "Progress in Inorganic Chemistry," Vol. 21,
      S. J. Lippard, ed., John Wiley and Sons, 1976, pp. 129-158.

(37) B. D. Cutworth, C. G. Davies, P. A. W. Dean, R. J. Gillespie,
      P. Ireland and P. K. Ummat, Inorg. Chem., 12, 1343 (1973).

(38) B. D. Cutworth, R. J. Gillespie and P. R. Ireland, Chem.
      Commun., 723 (1973).

(39) I. D. Brown, B. D. Cutworth, C. G. Davies, R. J. Gillespie,
      P. Ireland and J. E. Vekris, Can. J. Chem., 52, 791 (1974).

(40) P. A. W. Dean and R. J. Gillespie, Chem. Commun., 853 (1970).

(41) R. J. Gillespie and O. C. Vaidya, ibid., 40 (1972).

(42) J. B. Beal, Jr., C. Pupp and W. E. White, Inorg. Chem., 8,
      828 (1969).

(43) D. E. McKee and N. Bartlett, ibid., 12, 2738 (1973).

(44) Z. K. Nikitina and V. Ya. Rosolovskii, Izv. Akad. Nauk SSSR,
      Ser. Khim., 2173 (1970).

(45) N. J. Bjerrum's chapter, this volume.

(46) R. J. Gillespie and P. K. Ummat, Inorg. Chem., 11, 1674
      (1972).

(47) R. J. Gillespie, J. Passmore, P. K. Ummat, and O. C. Vaidya,
      ibid., 10, 1327 (1971).

(48) H. S. Low and R. A. Beaudet, J. Amer. Chem. Soc., 98, 3849
      (1976).

(49) J. Barr, D. B. Crump, R. J. Gillespie, R. Kapoor, and
      P. K. Ummat, Can. J. Chem., 46, 3607 (1968).

(50) R. J. Gillespie and P. K. Ummat, ibid., 48, 1239 (1970).

(51) J. Barr, R. J. Gillespie, K. Kapoor, and G. P. Pez, J. Amer.
      Chem. Soc., 90, 6855 (1968).

(52) J. Barr, R. J. Gillespie, G. P. Pez, P. K. Ummat, and
      O. C. Vaidya, Inorg. Chem., 10, 362 (1971).

(53) R. J. Gillespie and M. J. Morton, ibid., 9, 811 (1970).

(54) R. J. Gillespie and M. J. Morton, ibid., 11, 586 (1972).

(55) O. Glemser and A. Smalc, Angew. Chem., Intern. Ed., 8, 517
      (1969).

(56) A. J. Edwards and G. R. Jones, J. Chem. Soc., A, 2318 (1971).

(57) R. J. Gillespie and J. B. Milne, Inorg. Chem., 5, 1577
      (1966).

(58) C. Davies, R. J. Gillespie, and J. M. Sowa, Can. J. Chem.,
      52, 791 (1974).

(59) J. A. Plambeck, "Fused Salt Systems," Vol. X of "Encyclopedia
      of Electrochemistry of the Elements," A. J. Bard, ed.,
      M. Dekker, New York, 1976.

(60) K. W. Fung and G. Mamantov, in "Advances in Molten Salt
      Chemistry," Vol. 2, J. Braunstein, G. Mamantov, and G. P.
      Smith, eds., Plenum Press, 1973, pp. 199-254.

(61) K. A. Paulsen and R. A. Osteryoung, J. Amer. Chem. Soc., 98,
      6866 (1976).

(62) R. Marassi, G. Mamantov and J. Q. Chambers, Inorg. and Nucl.
      Chem. Letters, 11, 245 (1975).

(63) R. Marassi, G. Mamantov and J. Q. Chambers, J. Electrochem.
      Soc., 123, 1128 (1976).

(64) G. Mamantov, R. Marassi, J. P. Wiaux, S. E. Springer and
      E. J. Frazer, paper in preparation.

(65) S. E. Springer, R. Marassi, R. Huglen, G. Mamantov,
      N. R. Smyrl, F. W. Poulsen and N. J. Bjerrum, paper in
      preparation.

(66) H. E. Doorenbos, J. C. Evans and R. O. Kagel, J. Phys.
      Chem., 74, 3385 (1970).

(67) J. Robinson, B. Gilbert and R. A. Osteryoung, Inorg. Chem.,
      in press.

(68) B. Tremillon, A. Bermond and R. Molina, J. Electroanal.
      Chem., 74, 53 (1976).

(69) P. Hagenmuller, J. Rouxel, J. David, A. Colin and
      B. LeNeindre, Z. Anorg. Allg. Chem., 323, 1 (1963).

(70) P. Palvadeau and J. Rouxel, Bull. Soc. Chim. Fr., 2698
      (1967).

(71) J. Rouxel and P. Palvadeau, ibid., 2044 (1976).

(72) B. Gilbert and R. A. Osteryoung, to be submitted for publication.

(73) J. Robinson and R. A. Osteryoung, unpublished results.

(74) K. W. Fung and G. Mamantov, J. Electroanal. Chem., 35, 27 (1972).

(75) B. Gilbert, G. Mamantov and K. W. Fung, Inorg. Chem., 14, 1802 (1975).

(76) G. J. Hills, D. J. Schiffrin and J. Thompson, Electrochim. Acta, 19, 657, 671 (1974).

(77) G. J. Hills, D. J. Schiffrin and J. Thompson, J. Electrochem. Soc., 120, 157 (1973).

(78) P. Rolland and G. Mamantov, unpublished work.

(79) I. V. Vinarov, Russ. Chem. Rev., 36, 522 (1967).

(80) R. Huglen and G. Mamantov, unpublished work.

(81) G. Ting, K. W. Fung, and G. Mamantov, J. Electrochem. Soc., 123, 624 (1976).

(82) L. E. McCurry and G. Mamantov, unpublished work.

(83) G. Ting, Ph.D. Dissertation, The University of Tennessee, August 1973.

(84) L. E. McCurry, G. Mamantov, N. J. Bjerrum, F. W. Poulsen and J. H. von Barner, paper to be presented at the Meeting of the Electrochemical Society, Atlanta, Ga., October 1977.

(85) J. Phillips and R. A. Osteryoung, J. Electrochem. Soc., in press.

(86) D. L. Brotherton, Ph.D. Dissertation, The University of Tennessee, December 1974.

(87) F. de Joode and G. Mamantov, unpublished work.

(88) G. Mamantov, in Molten Salts: Characterization and Analysis, G. Mamantov, ed., M. Dekker, New York, N.Y., 1969.

(89) S. Senderoff and G. W. Mellors, Science, 153, 1475 (1966).

(90)   K. Schulze and H. Hoff, Electrochim. Acta, 17, 119 (1972).

(91)   G. L. Holleck and J. Giner, J. Electrochem. Soc., 119, 1161
       (1972).

(92)   B. Gilbert, D. L. Brotherton, and G. Mamantov, ibid., 121,
       773 (1974).

(93)   R. J. Gale and R. A. Osteryoung, ibid., 121, 983 (1974).

(94)   P. Rolland and G. Mamantov, ibid., 123, 1299 (1976).

(95)   L. G. Boxall, H. L. Jones and R. A. Osteryoung, ibid., 121,
       212 (1974).

(96)   R. J. Gale and R. A. Osteryoung, Inorg. Chem., 14, 1232
       (1975).

(97)   R. Marassi, J. Q. Chambers and G. Mamantov, J. Electroanal.
       Chem., 69, 345 (1976).

(98)   H. L. Jones, L. G. Boxall and R. A. Osteryoung, ibid., 38,
       476 (1972).

(99)   H. L. Jones and R. A. Osteryoung, ibid., 49, 281 (1974).

(100) P. Delahay, "New Instrumental Methods in Electrochemistry,"
       Interscience, N.Y., 1954, p. 90.

(101) J. P. Saget and V. Plichon, Bull. Soc. Chim. Fr., 1395
       (1969).

(102) D. E. Bartak and R. A. Osteryoung, J. Electroanal. Chem.,
       74, 68 (1976).

(103) G. Mamantov, R. Marassi and J. Q. Chambers, "High Energy
       Cathodes for Fused Salt Batteries," Technical Report
       ECOM-0060-F, April 1974.; G. Mamantov, R. Marassi and
       J. Q. Chambers, Extended Abstract #8, Fall Meeting of the
       Electrochemical Society, New York, N.Y., October 13-17.
       1974.

(104) K. W. Fung, J. Q. Chambers, and G. Mamantov, J. Electroanal.
       Chem., 47, 81 (1973).

(105) G. C. Demitras and E. L. Muetterties, J. Amer. Chem. Soc.,
       99, 2796 (1977).

ELECTROCHEMICAL AND SPECTROSCOPIC STUDIES OF THE CHALCOGENS IN CHLOROALUMINATE MELTS

Niels J. Bjerrum

The Technical University of Denmark, Chemistry
Department A
DK-2800 Lyngby, Denmark

1. INTRODUCTION

The study of lower oxidation states of post-transition elements in molten salts is a fascinating area which has been enlarged considerably in recent years (1). The investigation of the homolog series of the chalcogens (S,Se,Te,Po) has proven a very interesting field. Many especially interesting chemical features of these elements involve their ability to catenate (i.e. form chains) and in certain cases to give ring structures.

Only the positive oxidation states of the chalcogens in chloroaluminate melts will be considered here. For historical reasons we will start with the heavy elements. Even the longest lived polonium isotope has a half life of ca. 100 years and is, therefore, very difficult to work with. Apparently no experiments have been performed in chloroaluminate melts. We will, therefore, begin with tellurium. So far six different oxidation states of tellurium have been identified in these melts. The species are the following: Te(IV), Te(II), Te_2^{2+}, Te_6^{2+}, and Te_8^{2+}. This is, of course, a formal way of presenting these species. We know from measurements with chlorine-chloride electrodes (2), spectrophotometric measurements (2), and Raman spectra (3) that in the case of tetravalent tellurium at least four chloro complexes, $TeCl_6^{2-}$, $TeCl_5^-$, $TeCl_4$, and $TeCl_3^+$, are formed. In the case of divalent tellurium probably two complexes, $TeCl_3^-$ and $TeCl_2$, are formed (4), whereas as far as we know the other oxidation states of tellurium do not form chloro complexes in the chloroaluminate melts.

The smaller units, Te(II) and Te_2^{2+}, have been found in chloroaluminate melts buffered with $KCl-ZnCl_2$ (5). Like all the low oxidation

states of tellurium, they are most easily made by reaction between
elemental tellurium and $TeCl_4$. Evidence for the existence of
these species is based on spectrophotometric examination of the
resulting melts. This includes examination of the equilibrium:
$3Te^{2+} \rightleftharpoons Te_2^{3+} + Te^{4+}$, and of heterogeneous equilibria involving
solid or liquid tellurium together with a species with a known
formal charge. Equilibria of this type have in the past been used
to identify the low oxidation states of bismuth (6,7) and to a
lesser extent the lower oxidation states of tellurium (5,8). Unfortunately, it is usually rather difficult to establish the
desired equilibria for heterogeneous reactions with the suitable
species concentration. The parameters which can be changed are
the temperature, pCl^-, and the activity of the metal phase. All
these things have been tried successfully (5-8). It was possible
to establish that one of the species, Te_2^{2+}, on the high chloride
activity side had the charge 2+. Using this species as the species
with the known formal charge it could be shown that the other
species, Te^{2+}, in the basic melt had the same formal charge as
Te_2^{2+}; the activity of the metal phase was in the latter case
lowered by alloying it with gold (5). It is perhaps noteworthy
that p-p transitions, analogous to those found for Bi^+ ions, are
possible for divalent tellurium (5). The evidence for existence of
Te_4^{2+} in solution (8) was based partly on an equilibrium between
solid tellurium, Te_2^{2+} and Te_4^{2+} (this gave the formal charge 2+),
and partly on the stoichiometric reaction: $7Te + Te^{4+} = 2Te_4^{2+}$. As
mentioned above, it is not always possible to obtain the charges
on the unknown species from heterogeneous reactions. This can now
be done by a new combination of methods (9). This approach is based
on "model discrimination" using spectrophotometric and potentiometric measurements and will be described in some detail later (see
"Results and Discussion for the Tellurium Systems").

So far six different oxidation states of selenium have been
identified in acidic chloroaluminate melts: Se(IV), (Se_2^{2+}), Se_4^{2+},
Se_8^{2+}, (Se_{12}^{2+}), (Se_{16}^{2+}) (10). The parentheses indicate that the
formulas are tentative. In a recently published work based on
spectrophotometric measurements four species with low oxidation
states (all except (Se_2^{2+}) were found (11). Only the oxidation
states for two of these species could be determined (+1/2 and +1/4)
whereas the charge could not be determined for any of them. The
spectra based on the given formulas were also calculated (11).

In the case of sulfur, spectrophotometric (12) and potentiometric (13) measurements made by us in $NaCl-AlCl_3$ melts indicate
formation of S(IV), S(II), S_2^{2+}, S_4^{2+}, S_8^{2+}, and S_{16}^{2+}. At present not
much has been published about the positive oxidation states of
sulfur in chloroaluminate melts. On the basis of cyclic and pulse
violtammograms for solutions of sulfur in $NaCl-AlCl_3$ melts, Marassi,
Mamantov, and Chambers (14) assumed formation of S(I) and S(II)
by oxidation (as S^{2-} by reduction).

The existence of the solute chalcogen species in chloroaluminate melts is not unexpected in view of the results obtained for
the solid phases. Also, as is usually the case, more species can
be formed in the liquid than in the solid state. This is illustrated in the case of tellurium by the fact that for only one of
the species (Te_4^{2+}) found in solution have crystal structures been
described (Te_4^{2+}, with either $AlCl_4^-$ or $Al_2Cl_7^-$ as anion) (15). A
second tellurium species has also been obtained in the solid phase
(with oxidation state +1/3) (16). However, it has not been possible to obtain the X-ray structure.

In a study of the system, $SeCl_4$ + $4AlCl_3$ + xSe, only two
selenium species were found. These species were Se_4^{2+} and Se_8^{2+} both
with $AlCl_4^-$ as anion (16). Only for $Se_8(AlCl_4)_2$ has a crystal
structure been published (17).

It seems very difficult, if not impossible, to obtain any
crystals containing the low oxidation states of sulfur together
with chloroaluminate anions. By cooling the melt the elemental
sulfur is recovered (1).

Finally for the sake of completeness it should be mentioned
that solvents other than molten salts can be used to form low
oxidation states of the chalcogens. In fact, this is the tradi-
tional way of preparing some of the low oxidation states, although
it is only rather recently that their true nature has been revealed
(for review see Gillespie and Passmore) (18,19). The principal
solvents used are HSO_3F, $H_2S_2O_7$, SO_2 + SbF_5, and SO_2 + AsF_5. One
disadvantage of these solvents is that the solutions formed are
not always stable, since the species are oxidized by the solvent.
In contrast to the situation in chloroaluminate melts, it is especially important that it is not possible to get the species in
equilibria with each other even if stable to oxidation by the
solvent. However, it is possible to stabilize some of the low
oxidation states of the chalcogens (which cannot be stabilized by
$AlCl_4^-$ in the solid state) in crystals with AsF_6^-, SbF_6^-, $Sb_2F_{11}^-$,
SO_3F^- as anions. The species S_{16}^{2+}, S_8^{2+}, and S_4^{2+} have been isolated
in this way (18,19).

2.   EXPERIMENTAL

The preparation of the materials used ($NaCl$, $AlCl_3$, $TeCl_4$,
and $SeCl_4$) and the experimental techniques used for the tellurium
and selenium work have been described elsewhere (5,9,11). Since
none of the sulfur chlorides have a sufficiently low vapor pressure
at room temperature to be weighed in a glove-box prior to addition
to the melt, the sulfur species were produced electrolytically.
The cell for spectrophotometric measurements was made entirely of

fused quartz and consisted of a cuvette with an anode of glassy
carbon fused into the top. This cell was made small enough to fit
into our normal spectrophotometric furnace (11). The cathode was
made of platinum. For both electrodes the connection through the
fused quartz was made of molybdenum. The anode and the cathode
chambers were separated by a porous plug consisting primarily of
silica. When the dissolved sulfur (sulfur has a fairly high
solubility in the melt used) was to be oxidized, the cell was
turned upside down and electrical contact thus made between anode
chamber and cathode chamber through the porous plug. The desired
number of coulombs was then passed through the system by means of
a chronoamperostat (20) and the cell again turned to the upright
position. A special cell was also necessary for the potentiometric
measurements. This Pyrex cell had three chambers connected in
pairs by three capillaries fitted with porous plugs. The third
chamber was necessary in order to electrolytically change the
composition in the two main chambers independently. A detailed
description will be published shortly (12).

3. RESULTS AND DISCUSSION FOR THE TELLURIUM SYSTEMS

Three low oxidation states of tellurium, Te_4^{2+}, Te_6^{2+}, Te_8^{2+},
exist in the low melting NaCl-AlCl$_3$ (37:63 mol %) melt. The
existence of Te_4^{2+} in the melt has been proved earlier (8). The
tellurium species other than Te_4^{2+} can be found from "model
discrimination" performed on equilibria of the type

$$hTe_q^{r+} + jTe_y^{z+} \rightleftarrows kTe_v^{x+} \tag{1}$$

where h, j, and k are, of course, all integers. The measurements
are spectrophotometric and potentiometric. The spectrophotometric
method is based on Bouguer-Beer's law and the law of additive
absorbances. From this a general equation involving three matrices
can be put forward

$$[\ell_m C_{mi}][\varepsilon_i(\nu_n')] = [A_m(\nu_n')] \tag{2}$$

Methods of treating this matrix equation have been given by several
authors (21-23). Based on a particular chosen reaction scheme and
an arbitrarily chosen equilibrium constant the concentration of
each model species in a particular chosen reaction can be calculated
for each composition examined (11). The calculated concentration
for each of the three species can then be inserted in Eq. 2 together
with the measured absorbances obtained at many different wavenumbers. Eq. 2 cannot be solved exactly (it is overdetermined). But
it can fairly easily be solved in such a way that the "best"
solution (least square) is obtained and the deviation between the

measured and calculated absorbances can be found.   This is most
conveniently expressed in the form of variances.   The best values
of the integers in Eq. 1 are then found by systematic variation.
To simplify the analysis, the charge on a species was limited to
the likely maximum value of 4+ (9).

The best way to determine the number of species in the systems
spectrophotometrically is to plot at different wavenumbers the
formal absorptivity of one of the added substances versus the
formality ratio of the added substances, in the present case
$Te/TeCl_4$.   This ratio and the similar ratio for selenium, $Se/SeCl_4$,
will for convenience be designated R.   It can be shown that straight
lines on such a plot which are independent of the total concentration are an indication of a two-species system (9).   If this rule
is applied while examining the different peaks in the spectra (due
to the different species in the melt), it is possible to decide how
many species are present in the melt.   Furthermore, it is possible
to conclude whether more than three species are present (in
measurable amounts) at the same composition in the non-linear
ranges.

As a demonstration of the described principles a plot of the
formal absorptivity of $TeCl_4$ versus the formality ratio R is shown
in Figure 1.   A clear two-species (linear) range between R=0 and
R=ca.7 is seen here.   The variances for the spectrophotometric measurements for models with oxidation states close to +1/2 are given
in Table I.   The values marked with asterisks are within the 90%
probability range (based on an F-test) of the lowest obtained
variances (9,11).   As a general rule one can say that the spectrophotometric method is superior to the potentiometric method for
identifying different species in the melt, whereas the potentiometric method is superior in accuracy and hence for the purpose of
distinguishing between the different models.   The spectrophotometric measurements can give no information in the linear ranges
concerning the charge on the species.   However, considerably more
information can be obtained potentiometrically, when a two-chamber
redox concentration cell is used.

If a redox pair is being examined

$$qTe_v^{x+} \rightleftarrows vTe_q^{r+} + ne^-$$ (3)

where e^- is the electron and n the number of electrons, the
measured emf can be shown to be (9)

$$\Delta E = \frac{1}{F} \int_I^{II} \Sigma_i t_i (-z_i^{-1} d_{\mu_i} + n^{-1} vd_{\mu_{Te_q^{r+}}} - n^{-1} qd_{\mu_{Te_v^{x+}}})$$ (4)

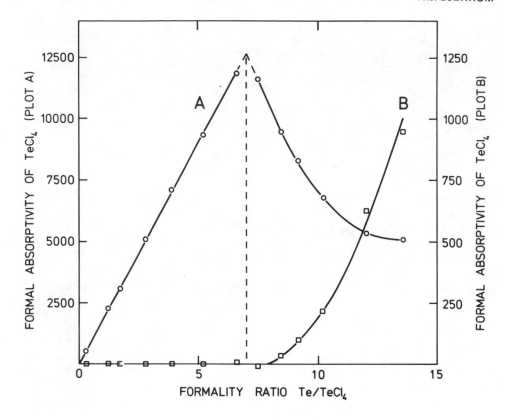

Figure 1.   Relation between the formal absorptivity of TeCl$_4$ measured at two different wavenumbers and the Te:TeCl$_4$ formality ratio (Ref. 9).

where t_i, z_i, and μ_i are the transference number, charge, and chemical potential respectively, for the ion numbered i, and roman numerals refer to cell chambers. It can be shown (9) that well within the limits of other experimental error ΔE can be expressed as

$$\Delta E = \frac{RT}{F} \ln \frac{[Te_q^{r+}]_{II}^{n^{-1}v} [Te_v^{x+}]_I^{n^{-1}q}}{[Te_q^{r+}]_I^{n^{-1}v} [Te_v^{x+}]_{II}^{n^{-1}q}} \tag{5}$$

Table I

Variances Based on Spectrophotometric Measurements for
Models[a] with Oxidation States Close to +1/2[b]

Oxidation state	Formal Charge			
	1+	2+	3+	4+
+2/3		1.3×10^{-3}		1.1×10^{-3}
+3/5			6.0×10^{-4}	
+4/7				7.0×10^{-4}
+1/2	$1.8 \times 10^{-5*}$	$1.8 \times 10^{-5*}$	$2.7 \times 10^{-5*}$	$1.9 \times 10^{-5*}$
+4/9				2.3×10^{-3}
+3/7			5.2×10^{-4}	
+2/5		1.7×10^{-1}		1.5×10^{-3}

[a]Other species in the equilibria assumed to be Te^{4+} and Te_6^{2+}.
[b]Formality ratio range 0.3 - 9.2.

where R, T, and F have their usual meaning. If the composition in
cell chamber I is kept constant, ΔE plotted against $\log([Te_{II}^{r+}]_{II}^{n-1}v/[Te_V^{x+}]_{II}^{n-1}q)$ should give a straight line with the slope RT $(\ln 10)/F$
if the correct model is being considered. If the correct
model has not been selected, curved lines with a "slope" different
from RT ln 10/F will result. Unfortunately in the case of sulfur,
selenium, and tellurium (and for some of the other elements which
can form low oxidation states) not enough linear ranges are present
to determine the formulas for all the species involved. It should,
furthermore, be noted that within a truly linear range equilibrium
constants are meaningless. It is, therefore, necessary also to
consider non-linear ranges.

A model discrimination can also be made using the potentiometric measurements (Figure 2) if a three-species range is
being examined. As in the spectrophotometric measurements, one
starts with Eq. 1. The concentrations of the three species are
then calculated and inserted in Eq. 5. By varying systematically the equilibrium constants for the particular model equilibria
considered, it is possible to minimize the deviation (least
square) between the formality ratios calculated from ΔE and

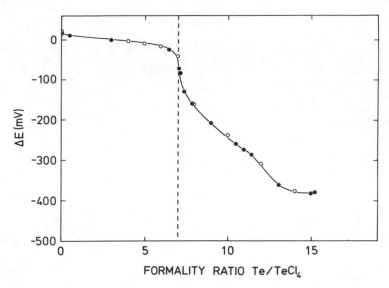

Figure 2.   Measured ΔE values are a function of the Te: TeCl$_4$ formality ratio (Ref. 9).

the formality ratios calculated directly from the weighings.   By varying systematically the different models, it is possible to obtain a table of the variances for the different models (Table II) calculated for approximately the same R-range as the one obtained for the spectrophotometric measurements (Table I).   A comparison between these tables results in a single model, Te$_4^{2+}$, within 90% probability.   The formulas for the other tellurium species can be found in a similar way (9).

In conjunction with solving the matrix equation, Eq. 2, the spectra of the individual species are obtained.   In Figure 3 are given the calculated spectra of the 4 different tellurium species found in the acidic NaCl-AlCl$_3$ melt.   Finally in Table III are given the pK values for the different tellurium equilibria calculated by both methods, together with the non-linear confidence limits.   It is not possible to compare the two methods directly, since they were not performed at the same temperature.   However, there seems to be no contradiction between the two methods if the temperature is taken into consideration.

4.   RESULTS AND DISCUSSION FOR THE SELENIUM SYSTEMS

In the low melting NaCl-AlCl$_3$ melt (37:63 mol %) two selenium species with low oxidation states have been identified with reasonable certainty.   These are Se$_4^{2+}$ and Se$_8^{2+}$.   Three other low

Table II

Variances Based on Potentiometric Measurements for Models[a]
With Oxidation States Close to $+1/2$[b]

| Oxidation state | Formal Charge | | | |
	1+	2+	3+	4+
+2/3		(1.3×10^4)		(4.4×10^3)
+3/5		$(2.2 \times 10^{3*})$		
+4/7				(3.7×10^3)
+1/2	3.5×10^3	$1.1 \times 10^{3*}$	(3.7×10^3)	4.9×10^3
+4/9				(5.5×10^3)
+3/7			(6.2×10^3)	
+2/5		4.4×10^3		5.9×10^3

[a] Other species in the equilibria assumed to be Te^{4+} and Te_6^{2+}.
[b] Formality ratio range 0.0 - 7.0.

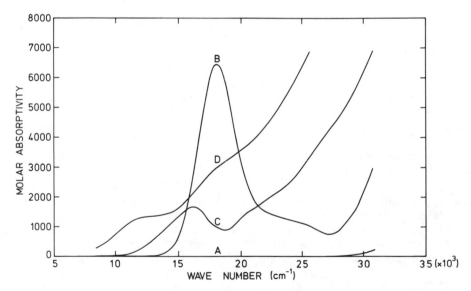

Figure 3. Calculated spectra of (A) Te^{4+} (mainly as $TeCl_3^+$), (B) Te_4^{2+}, (C) Te_6^{2+}; and (D) Te_8^{2+} (Ref. 9).

Table III

pK Values[a] for the Equilibria Between Tellurium Species in
Solution in Molten $NaCl-AlCl_3$ (37:63 mol %)
At 150° and 250°C

Equilibrium	Method			Confidence limits for pK	
$2Te^{4+} + 7Te_6^{2+} \rightleftarrows 11Te_4^{2+}$	spec.	250°C	-12	-∞	to -9
	poten.	150°C	-17	-∞	to -10
$Te_4^{2+} + Te_8^{2+} \rightleftarrows 2Te_6^{2+}$	spec.	250°C	-0.4	-2.2	to -0.2
	poten.	150°C	-0.93	-1.03	to -0.84

[a]Based on molar concentrations.

oxidation states, assumed to be SE_2^{2+}, Se_{12}^{2+} and SE_{16}^{2+}, have also been found. The experimental methods (spectrophotometric and potentio-metric) are similar to those described for tellurium.

An examination of the formal absorptivity of $SeCl_4$ versus the formality ratio Se:$SeCl_4$ (11) gave valuable information. As can be seen from Figure 4 the formality ratio range in which $SeCl_4$ can be reduced by elemental selenium is much broader (R-range 0 to ca. 31) than is the case for the tellurium system (Figure 1) (R-range 0 to ca. 15). As for the tellurium system a sharp break is observed for R = 7. A less marked break is observed around R = 15, whereas no well defined breaks are observed in the R-range 15 to 31. Almost linear ranges are observed between 0 and 7 and between 7 and 11. The spectra indicates that outside these limits three species ranges are present. A calculation performed on the measured spectra using the matrix equation (Eq. 2) given in connection with the tellurium work, showed that the spectra in the formality ratio range 1.3 to 11.1 could be described best by an equilibrium involving Se(IV), a selenium species with oxidation state +1/2, and a selenium species with oxidation state below +1/2. This selenium species with oxidation state below +1/2 was by calculations in the R-range 9.0 to 21.3 found to have the oxidation state +1/4 (or less likely +4/17). For the R-range 21.3 to 32.0 the oxidation states of two species with oxidation states below +1/4 could not be determined. However, the most likely values for the oxidation states of these two species were found to be +1/6 and +1/8.

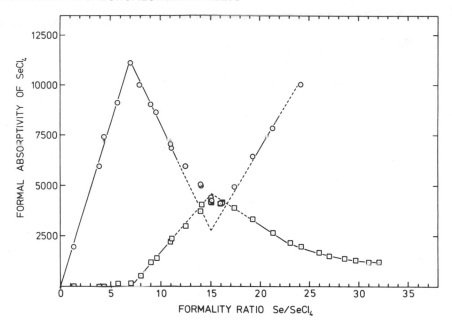

Figure 4. Relation between the formal absorptivity of SeCl$_4$ measured at two different wavenumbers and the Se:SeCl$_4$ formality ratio (Ref. 11).

In Figure 5 is shown the result of some potentiometric measurements of the same type as described for tellurium. The
measured potentials of a two-chamber concentration cell are plotted
versus the formality ratio Se/SeCl$_4$. If this plot is compared with
the similar plot for the tellurium system (Figure 2), it can be
seen that there are more inflection points on the selenium plot
than on the tellurium plot. This is in agreement with the spectrophotometric measurements, which indicate that more species are
present in the selenium-containing melts than in the similar
tellurium-containing melts. It can be seen that there is an
inflection point around R = 7 but this is not nearly as pronounced
as in the tellurium-containing system. In full agreement with the
spectrophotometric measurements there is a pronounced change
(inflection point) around the formality ratio 15. Furthermore,
there is one more inflection point in the R-range 15 to 31. Above
an R-value of ca. 31 there is no change in the measured potential
indicating that the melt is now saturated with selenium. As ex-
plained for tellurium, ΔE plotted versus $\log([X_q^{r+}]_{II}^{n-1}v/[X_v^{x+}]_{II}^{n-1}q)$
(X = S, Se, and Te) should give a straight line with the slope RT
(ln 10)/F for a two-species range. Such a plot is shown for the
selenium-containing melt for the R-range 1.0 to 5.0 (Figure 6).
Four species with the general formulas Se$_{2n}^{n+}$ (n = 1,2,3,4) are
compared here. The other species, of course, involves Se(IV).

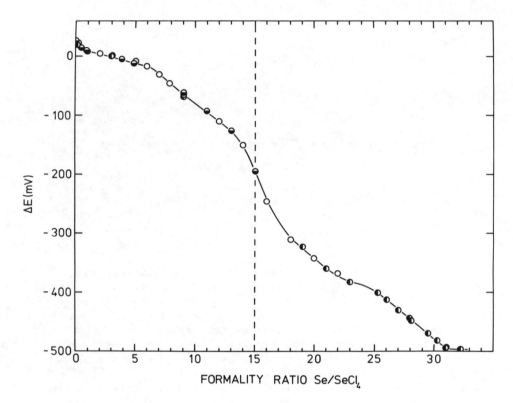

Figure 5.   Measured ΔE values as a function of the Se:SeCl$_4$ formality ratio.

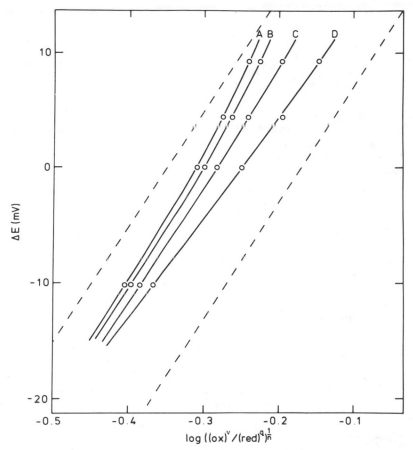

Figure 6. Relation between the measured ΔE and the calculated value of $\log((Ox)^v/(Red)^q)^{1/n}$ for the formality ratio range 1.0 to 5.0. Formulas for Ox and Red: A, Se^{4+}, Se_8^{4+}; B, Se^{4+}, Se_6^{3+}; C, Se^{4+}, Se_4^{2+}; D, Se^{4+}, Se_2^+. The dashed lines have the theoretical slope.

None of the lines in Figure 6 are straight. However, the plot C, which represents Se_4^{2+}, although not entirely linear, has a slope closest to the theoretical value. This is rather surprising since in the same range for the tellurium-containing system a straight line with the correct slope is obtained for the species Te_4^{2+}. This peculiar problem now seems to be solved. A closer spectrophotometric examination (i.e. more compositions examined at more wavenumbers) in the R-range 0 to 7, provided evidence for the presence of a small amount of a new species. The most likely species would seem to be Se_2^{2+}. This is based partly on the measurements and partly on the fact that the corresponding tellurium species Te_2^{2+} is

found in basic and neutral $KAlCl_4$ melts (5). Calculations allowing for the possible presence of this species and based on the
potentiometric measurements gave corresponding results. The calculations based on the best fit (least squares) indicated that
this species had only a rather small maximum concentration and was
of importance in only a rather small R-range (10).

Model discrimination using the potentiometric measurements for
the range 3.0 to 12.0 combined with the spectrophotometric measurements for a similar range, gave Se_4^{2+} as the most likely species of
next lower oxidation state (75% probability (10)). If we look at
the models with oxidation state close to +1/4 it is found that
only two species, Se_8^{2+} and Se_{11}^{3+}, are within the 90% probability
limit. Se_{11}^{3+} is excluded by comparison with the results from the
spectrophotometric measurements, leaving only Se_8^{2+}. Calculations
on equilibria involving the species with oxidation states lower
than +1/4 are, as mentioned above, not very discriminative. However, based on the spectrophotometric measurements, the oxidation
state +1/6 seems the most likely, especially in the form of Se_{12}^{2+}.
For the next oxidation state many formulations are possible, but
the most likely possibility is Se_{16}^{2+}. It is interesting to compare
the equilibrium constants for all the described reactions. This is
done in Table IV. From this Table it can be seen that where
comparison is possible there is good agreement between the pK-
values obtained by the two methods.

As mentioned in the "Introduction" chloro complexes seem to
be found only for the higher oxidation states of the chalcogens.
In the case of tetravalent selenium only one chloro complex seems
to be important in the $NaCl-AlCl_3$ molten salt system. This chloro
complex is $SeCl_3^+$. This conclusion is based on the following results. No change is observed in the Raman spectra due to variation
in the acidity/basicity of the melt using 0.2 and 0.5 F solutions
of $SeCl_4$ in $NaCl-AlCl_3$ at 175°C. Changes are, of course, found for
the $NaCl-AlCl_3$ solvent (24). The intensity of the bands due to the
selenium complex is proportional to the concentration of the tetravalent selenium (comparison is made with solvent bands). All this
indicates only one important selenium complex. This assumption is
in complete agreement with the potentiometric measurements performed with chlorine/chloride electrodes in a concentration cell
described earlier (2,25). $\bar{n}$ (the average experimental coordination
number) is here found to be ca. 3 in most of the basic and most of
the acidic range except at low and high pCl^-, where $\bar{n}$ values above
and below 3 are found, corresponding probably to partial conversion
to $(SeCl_4)$ and $SeCl_2^{2+}$, respectively (26).

Table IV

pK Values[a] for the Equilibria Between Selenium Species
In Solution in Molten NaCl-AlCl$_3$ (37:63 mol %)
At 150°C

Equilibrium	Method	pK	Confidence limits for pK
$2Se^{4+} + 3Se_4^{2+} \rightleftarrows 7Se_2^{2+}$	poten.	7.9	4.7 to 9.1
$4Se^{4+} + 7Se_8^{2+} \rightleftarrows 15Se_4^{2+}$	poten. spec.	-12.5 b	$-\infty$ to -12.4 $-\infty$ to -5.5
$Se_4^{2+} + Se_{12}^{2+} \rightleftarrows 2Se_8^{2+}$	poten. spec.	-2.9 -2.0	$-\infty$ to -2.5 $-\infty$ to -0.7
$Se_8^{2+} + Se_{16}^{2+} \rightleftarrows 2Se_{12}^{2+}$	poten. spec.	0.0 -0.8	-0.2 to 0.2 $-\infty$ to 2.4

[a]Based on molar concentrations
[b]No minimum.

5.   RESULTS AND DISCUSSION FOR THE SULFUR SYSTEMS

In the case of sulfur, the spectrophotometric measurements
made so far indicate formation of S_2^{2+}, S_4^{2+}, S_8^{2+}, and S_{16}^{2+} together
with S(II) and S(IV), as the result of an anodic oxidation of
elemental sulfur in the low melting NaCl-AlCl$_3$ (37:63 mol %) at
150°C; all formulas are tentative.   In Figures 7A and 7B are shown
the spectra for the S/Cl$_2$ formality ratio ranges ∞ to 26.5 and 21.7
to 11.5, respectively.   Spectra for the range 11.5 to 0.6 are not
shown.   The production of the sulfur species is described in the
Experimental section.   The absorbance at one wavenumber as a function of the Cl$_2$/S ratio is shown in Figure 8.   As seen from this
curve there is a maximum near Cl$_2$/S = 0.063 corresponding to the
formal oxidation state +1/8 and a break at Cl$_2$/S = ca. 0.125
corresponding to +1/4.   As before, straight lines on such a plot
are taken as an indication that only two sulfur species are
present.   Therefore in the range 0 to 0.06, elemental sulfur and a
sulfur species with the oxidation state +1/8 are present.   In the
range 0.07 to 0.10 sulfur species with the oxidation states +1/8
and +1/4 are present and in the range 0.10 to 0.175 sulfur species
with oxidation states +1/8 and +1/4 are present, together with a
new oxidation state.   Another way to treat the measured absorbances

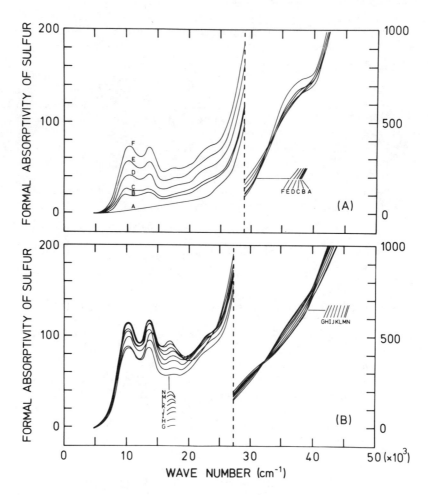

Figure 7.   Spectra of mixtures of different sulfur species in
NaCl-AlCl$_3$ (37:63 mol %) at 150°C.   Formality ratio S/Cl$_2$:   A,
"∞"; B, 115.1; C, 77.7; D, 47.3; E, 34.0; F, 26.5; G, 21.7; H,
19.4; I, 17.1; J, 16.0; K, 14.9; L, 13.7; M, 12.6; N, 11.5.

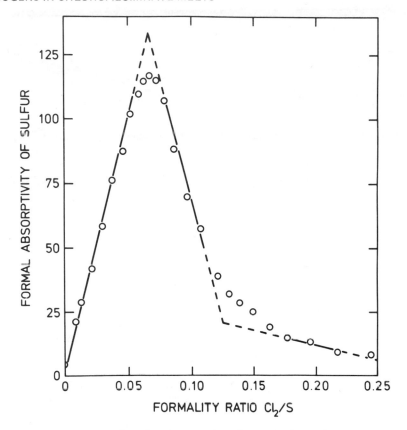

Figure 8.   Relation between the formal absorptivity of sulfur measured at 10.4×10^3 cm^{-1} and the Cl_2:S formality ratio.

is to plot them versus the formality ratio S/Cl_2 (Figure 9).   A break due to one of the two oxidation states mentioned above (+1/4) can be seen here.   The break due to the disappearance of the other oxidation state (+1/8) is found outside the depicted range.   Three new breaks are found at formality ratios of approximately 1/2, 1, and 4 corresponding to the oxidation states +4, +2, and +1/2, respectively.   The break near formality ratio 1 is not so well defined, and it can be seen that it may even be located at formality ratio 2 (corresponding to the oxidation state +1).   However, some earlier and preliminary results for 300°C clearly indicate a break at a ratio of 1.

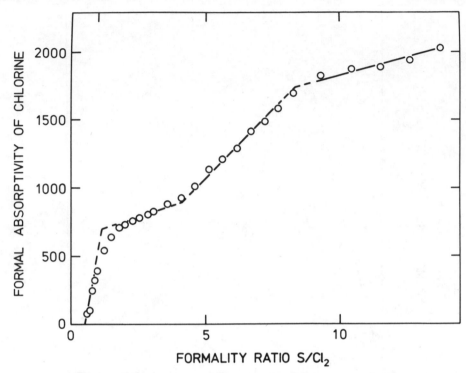

Figure 9.   Relation between the formal absorptivity of Cl_2 measured at 26.3×10^3 cm^{-1} and the $S{:}Cl_2$ formality ratio.

It would, of course, be useful to know the formal charges of these sulfur species. By comparison with the selenium and tellurium systems and the literature (18,19), the most likely charges seem to be 2+ (and 4+) and the species: elemental sulfur, S_{16}^{2+}, S_8^{2+}, S_4^{2+}, S_2^{2+}, S^{2+}, and S^{4+}. As in the case of tellurium and selenium it should be possible to determine this charge by potentiometric measurements. The potentiometric cell is described under "Experimental".

The measured potentials from this cell are shown in Figure 10. As can be seen, no abrupt changes take place (as in the case of selenium and tellurium), but the formation of the oxidation states +1/8 and +1/2 can be detected (from inflection points). The same methods applied to the potentiometric measurements on selenium and tellurium have been applied to the sulfur case. The calculations are, however, not yet completed. The most interesting results can be summarized as follows. The calculations indicate that it is necessary to include monovalent sulfur in the form of S_2^{2+} in equilibrium with tetravalent and divalent sulfur. By analogy with what is known about selenium and tellurium (2,26), S(IV) is assumed

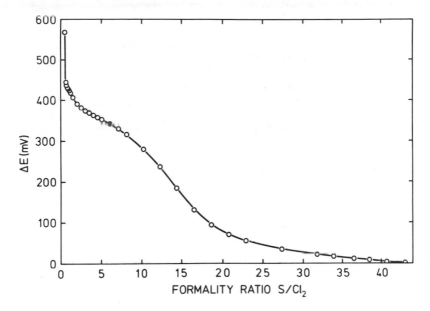

Figure 10.   Measured ΔE values as a function of the $S:Cl_2$ formality ratio.

to be present as SCl_3^+. The calculations show that the oxidation state +1/8 most probably occurs as S_{16}^{2+}. The best fit is obtained with the assumption that elemental sulfur is present as mostly S_8.

6.   SUMMARY

We have found that the use of spectrophotometric and potentiometric measurements, combined with a special calculation technique, has proved very useful for the study of the chalcogens in chloroaluminate melts.

This combination of methods has revealed the existence of the solvated entities Te_4^{2+}, Te_6^{2+}, and Te_8^{2+} in a low melting $NaCl-AlCl_3$ (37:63 mol %) solvent. The spectra of the individual tellurium species were calculated from the spectra of the experimental mixtures. pK values for the reactions $2Te^{4+}(soln) + 7Te_6^{2+}(soln) \rightleftarrows 11Te_4^{2+}(soln)$ and $Te_4^{2+}(soln) + Te_8^{2+}(soln) \rightleftarrows 2Te_6^{2+}(soln)$ were found from both types of measurements.

Spectrophotometric and potentiometric measurements on the solvated selenium species in the same solvent indicated the existence of Se_4^{2+} and Se_8^{2+}, and that Se_2^{2+}, Se_{12}^{2+}, and Se_{16}^{2+} were the most likely other low oxidation states formed by reduction of $Se(IV)$ with elemental selenium. pK values were calculated for the equilibria between these species. Raman spectroscopy and potentiometry on Se(IV) in NaCl-AlCl$_3$ showed that in most of the basic and in most of the acidic range Se(IV) was present as SeCl$_3^+$.

In the case of sulfur the low oxidation states in the NaCl-AlCl$_3$ melt were produced by anodic oxidation of elemental sulfur. The most consistent interpretation of the spectrophotometric and potentiometric measurements as well as information in the litera-ture was that S_2^{2+}, S_4^{2+}, S_8^{2+}, S_{16}^{2+} were formed together with S(II) and S(IV) (most likely as SCl$_3^+$).

7.   ACKNOWLEDGMENT

The author wishes to thank R. Fehrmann and J. C. Reeve for valuable discussions. This work was supported by a grant from Statens tekniskvidenskabelige Forskningsråd.

8.   REFERENCES

(1)   J. D. Corbett, Progr. Inorg. Chem., 21, 129 (1976).

(2)   J. H. von Barner, Niels J. Bjerrum, and K. Kiens, Inorg. Chem., 13, 1708 (1974).

(3)   F. W. Poulsen, N. J. Bjerrum, and O. F. Nielsen, Inorg. Chem., 13, 2693 (1974).

(4)   F. W. Poulsen and N. J. Bjerrum, J. Electroanal. Chem, in press.

(5)   N. J. Bjerrum, Inorg. Chem., 10, 2578 (1971); 11, 2648 (1972).

(6)   N. J. Bjerrum, C. R. Boston, and G. P. Smith, Inorg. Chem., 6, 1162 (1967).

(7)   N. J. Bjerrum and G. P. Smith, Inorg. Chem., 6, 1968 (1967).

(8)   N. J. Bjerrum, Inorg. Chem., 9, 1965 (1970).

(9)   R. Fehrmann, N. J. Bjerrum, and H. A. Andreasen, Inorg. Chem., 15, 2187 (1976).

(10) R. Fehrmann and N. J. Bjerrum, to be published.

(11) R. Fehrmann, N. J. Bjerrum, and H. A. Andreasen, Inorg. Chem., 14, 2259 (1975).

(12) R. Fehrmann, N. J. Bjerrum, and F. W. Poulsen, to be pub-
      lished.

(13) R. Fehrmann and N. J. Bjerrum, unpublished results.

(14) R. Marassi, G. Mamantov, and J. Q. Chambers, J. Electrochem.
      Soc., 123, 1128 (1976).

(15) T. W. Couch, D. A. Lokken, and J. D. Corbett, Inorg. Chem.,
      11, 357 (1972).

(16) D. J. Prince, J. D. Corbett, and B. Garbisch, Inorg. Chem.,
      9, 2731 (1970).

(17) R. K. McMullan, D. J. Prince, and J. D. Corbett, Inorg. Chem.,
      10, 1749 (1971).

(18) R. J. Gillespie and J. Passmore, Accts. Chem. Res., 4, 413
      (1971).

(19) R. J. Gillespie and J. Passmore, Chem. Britain 8, 475 (1972).

(20) H. A. Andreasen and N. J. Bjerrum, Inorg. Chem., 14, 1807
      (1975).

(21) Å. Björck, BIT 7, 1, 257 (1967).

(22) L. R. Lieto, Thesis; Report ORNL-TM-2714, Oak Ridge National
      Laboratory, Oak Ridge, Tenn., 1969.

(23) Z. Z. Hugus, Jr. and A. A. El-Awady, J. Phys. Chem., 75, 2954
      (1971).

(24) G. Torsi, G. Mamantov, and G. M. Begun, Inorg. Nucl. Chem.,
      Letters, 6, 553 (1970).

(25) J. H. von Barner and N. J. Bjerrum, Inorg. Chem., 12, 1891
      (1973).

(26) R. Fehrmann, J. H. von Barner, N. J. Bjerrum, and O. F.
      Nielsen, to be published.

ELECTRON FREE ENERGY LEVELS AND SPECTROSCOPIC CHARACTER OF DILUTE
SPECIES IN OXIDIC SOLVENTS: RELATING AQUEOUS TO LIQUID (AND
VITREOUS) OXIDE SOLUTIONS

C. A. Angell

Department of Chemistry, Purdue University

W. Lafayette, Indiana   47907

1.   INTRODUCTION

The familiar "basic" solution of aqueous solution chemistry
is produced by reacting the Lewis base Na_2O with the Lewis acid
H_2O initially to produce the "neutral" compound NaOH and by then
adding a large excess of the acid component (55.5 : 1 molar ratio
for a one molar solution). In an exactly analogous fashion an
ordinary silicate glass is produced by reacting the same Lewis
base Na_2O with the Lewis acid SiO_2, initially to form the neutral
Na_4SiO_6 compound, and by then adding a large excess of the acid
component.* It is not surprising then that the chemistry of the
two types of acid-base systems should have common features[†]:   both
are characterized by very low values of the mean oxide activity
$a_{Na_2O}^{\pm}$.

Related statements can be made for the oxidic acid-base systems Na_2O + CO_2 (which give us the molten carbonate electrolytes
of hydrocarbon-burning fuel cells) and for the other common
glass-forming oxide systems Na_2O + B_2O_3 or P_2O_5.

---

*For a "window" glass a certain amount of the weaker base CaO
is included for the very practical reason that the pure sodium
silicate solution is rather soluble in water.

†Interestingly enough, NaOH solutions of the same acid (H_2O)
to (Na_2O) ratio as the most studied Na_2O-SiO_2 solution, also have
glassforming ability and have been much used in low-temperature
radiation chemistry studies.

Physically, distinctions between the siliceous and aqueous systems arise in different ways. The most important difference results from differences in coordinating abilities of the positively charged components Si^{4+} and H^+ of the respective Lewis acids. In the case of SiO_2, both covalent and ionic bonding contributions require Si(IV) to be central to, and equally coordinated by, four oxygens. This results in a network of O-linked Si centers. In the case of H_2O, on the other hand, it is the oxygen which is coordinated by the positively charged species, and each oxygen is largely satisfied by the two covalently bonded protons of the isolated molecule. The "network" formed by the relatively weak coordination (through hydrogen bonds) of H^+ species on neighboring molecules is largely disrupted at ambient temperatures with the result that these acid-base systems are liquid at normal temperatures. The strong network of SiO_2, by contrast, is only degraded above 1200°C, and its fragments carry ionic charges so that siliceous systems only exhibit fluid behavior at very high temperatures. While this circumstance leads to gross physical differences between the two systems, the chemical similarity remains.

The different coordinating characteristics of H^+ and Si^{4+} may, however, be used to generate an essentially chemical difference between the two systems. The maximum coordination of oxygen by silicon is 2, and this is true also for all the other common inorganic Lewis acid cations (B^{3+}, etc.) except H^+. The oxygen of the OH_2 molecule can accept a third H^+. The addition of the third proton causes a further polarization of the oxygen and thus produces a further large drop in the thermodynamic activity of the oxide ion (in the vicinity of the (neutral) H_2O stoichiometry) for which no parallel exists in the siliceous systems.* It is, of course, in the vicinity of this stoichiometry that most aqueous chemistry is actually performed.

_____

*In the SiO_2 system, the nearest parallel of the autoionisation of H_2O

$$2H_2O \rightleftarrows H_3O^+ + OH^-$$

is the bond cleavage

$$-\overset{|}{\underset{|}{Si}}-O-\overset{|}{\underset{|}{Si}}- \rightleftarrows -\overset{|}{\underset{|}{Si}}^+ + \ ^-O-\overset{|}{\underset{|}{Si}}-.$$

The equilibrium constant is of the order $e^{-(10^5(cal)/RT)} \approx 2 \times 10^{-22}$ at 1000 K; $K_w = 10^{-14}$ at 298 K. There is no external source of $-\overset{|}{\underset{|}{Si}}^+$ by means of which this equilibrium can be displaced to reduce $[^-O-\overset{|}{\underset{|}{Si}}-]$.

With this outline of the chemical relationship between aqueous and other oxidic systems, let us point out similarities in the sort of solution chemistry problems which are faced by those interested in such solutions, and then outline a scheme within which problems involving changes of oxidation state can be treated equivalently for all systems.

Two important classes of problem which exercise both aqueous and glass chemists are (i) the assessment of structural arrangements of solvent species around solute species (e.g. how many OH_2, OH^- or $^-O-Si-$ species are there in the first coordination shell of V(III) Nd(III), etc.?) and (ii) the energetics of electron transfers between solute and solvent species, (e.g. will V(II) be stable in a given oxide medium?, will Fe occur as Fe(II) or Fe(III) in a given laser glass?). The answers to particular cases of each class prove to be rather sensitive to the value of $a_{\pm}^{oxide}$ in all oxidic media. Similarities between redox processes in aqueous and molten oxide systems, and between cation coordination schemes, have often been noted but no quantitative relationships have previously been established.

The main point of this chapter is to show that redox problems in all these oxidic media can be reduced to a common basis by organizing our thinking, and the experimental data from spectroscopy, from chemical analysis, and from electrochemistry, around the reference electron transfer process

$$\tfrac{1}{2} O_2^{2-} \text{ (solution)} \rightarrow \tfrac{1}{4} O_2 \text{ (gas, 1 atm)} + e^- \qquad (1)$$

which is common to all systems.

A secondary point is to demonstrate that the composition dependences of redox processes in oxidic media share a common feature. They are all, at first sight, anomalous, and can only be understood if it is assumed that electron transfers away from solute species do not occur as independent processes but rather occur in concert with transfer of oxide species (or at least oxide electron density) to the new higher oxidation number species. Physically this is to be recognized as an increased polarisation of the solvent medium by the higher oxidation number cation. Most of the ideas presented here have been developed in more detail in a discussion of the visible spectra of vitreous solids given elsewhere. (1)

2.   THE OXIDE-OXYGEN REFERENCE, ELECTRON TRANSFERS, AND EFEL DIAGRAMS FOR INDIVIDUAL SOLVENTS

In an oxidic solution in equilibrium with one atmosphere of oxygen, the distribution of a metallic element M between different valence states M^{n+} and $M^{(n+)+1}$ is determined by the reversible work ε expended to promote the transfer of an electron from the reduced state to the gaseous oxygen species, according to

$$K_{EQ} = e^{-\varepsilon/kT} \qquad (cf. N\varepsilon \equiv \Delta G = -RT \ln K_{EQ}) \qquad (2)$$

where K_{EQ} is the equilibrium constant for the exchange process

$$M^{n+} + \tfrac{1}{4} O_2 \text{ (g)} \rightleftarrows M^{(n+)+1} + \tfrac{1}{2} O^{2-} \qquad (3)$$

Normally the oxide ion will not be stable as such but will react either with a solvent species such as -Si-O-Si-, H_2O, etc., to give -Si-O$^-$, OH$^-$, etc. species, or with the oxidized species itself, but these are matters to be discussed later. Considering a solvent chosen to have unit activity of oxide and considering dilute solution standard states for the redox species such that activities can be replaced by mole fractions, K_{EQ} in Eq. (2) reduces to the concentration ratio of the two species. In such a solution the free energy of electron transfer can be simply determined from a chemical or spectroscopic analysis of the redox ratio. As Gurney (2) has emphasized, it is the quantity ε which is characteristic of the process; the equilibrium constant which we observe is only a quantity which, when taken as a logarithm and multiplied by -kT, provides a probabilistic (TdS) free energy change equal and opposite to ε thus permitting dG to be zero for small transfers. This, of course, is the condition which defines an equilibrium state (hence Eq. (2)). The quantity ε can be represented schematically by an analog of Gurney's diagram for proton transfer processes in aqueous solutions (2). Fig. 1 depicts ε as the interval between two levels in a diagram in which the oxidized and reduced states of the species are designated as "vacant" and "occupied" electron levels respectively. Thus we have for the ferrous-ferric equilibrium:

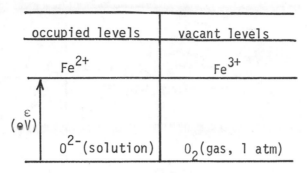

Figure 1

From Fig. 1 we can see that electrons from an excess of Fe^{2+} species will "fall out" into the lower-lying vacant levels on the O_2 species thereby forming O^{2-} (or oxygen in the -2 oxidation state) and leaving behind Fe^{3+}. The electron dumping will proceed until $RT\ln K = -\varepsilon$, i.e. until the excess has been eliminated.

If we choose the O^{2-}(soln) / O_2 (gas 1 atm) as a reference level, then the quantities ε for various electron transfer processes from reduced species to O_2 (or from O^{2-} to oxidized species) can be compared (i) with each other in the same solvent to establish a potential series in that solvent (ii) with themselves in other solvents to determine and interpret the influence of the medium on the electron transfer energy. Note that if the oxidic medium chosen were an Na_2O + H_2O solution of Na_2O content equivalent to ∿ one mole NaOH per kgm, the quantity ε in eV of Fig. 1 would be numerically equal to the aqueous solution standard oxidation potential E_B (volts) for oxygen evolution less the standard oxidation potential for the Fe(II)/Fe(III) electrode, i.e.

$$\varepsilon = -[E_B^{\,\circ} \text{ (oxygen evolution)} -E_B^{\,\circ} \text{ (Fe(II)/Fe(III))}]$$

This relationship is depicted in Fig. 2 which we refer to as an EFEL (Electron Free Energy Level) diagram.

In practice Fe(II) and Fe(III) are both precipitated from one molar NaOH solution as solid hydroxides. In the equivalent Na_2O + SiO_2 solution they remain in solution and can be also retained in the dissolved state in the glassy solid formed by cooling the melt to room temperature.

There are data of reasonable quality available for the concentration ratios K of a number of redox couples in the Na_2O + SiO_2 solvent of stoichiometry $Na_2O \cdot 2SiO_2$. These have been analyzed in

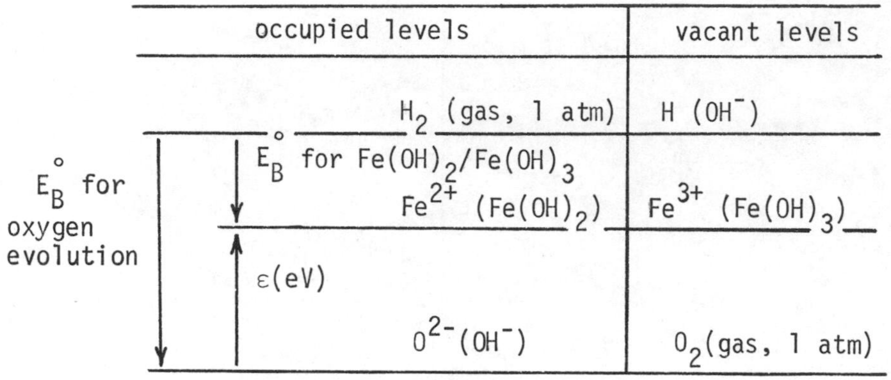

occupied levels	vacant levels

E_B^o for oxygen evolution

H_2 (gas, 1 atm)   H (OH⁻)

E_B^o for Fe(OH)$_2$/Fe(OH)$_3$

Fe^{2+} (Fe(OH)$_2$) Fe^{3+} (Fe(OH)$_3$)

ε(eV)

O^{2-}(OH⁻) O_2(gas, 1 atm)

Figure 2

some detail in Ref. 1 where an extensive comparison of the quantities ε derived using Eq. (1) has been made with their aqueous
solution equivalents obtained from Eq. (3) (except that the data
for the aqueous media were for one molar acid solutions in which
case there is no insolubility problem). To a lesser extent redox
ratios have been investigated in sodium borate solutions (3). In
Fig. 3 parts (a) (b) and (c) we show ε values for redox couples in
the solutions Na$_2$O· 112 H$_2$O (1 molar NaOH)(4), Na$_2$O·2SiO$_2$ (1), and
Na$_2$O·2B$_2$O$_3$ (1,3). For comparison, the ε values for standard
($\sim$ 1 molar) acid solutions, in which the oxide activity is 14
orders of magnitude lower than in the first of the above cases,
are shown in the right hand diagram, Fig. 3 (d). The energy level
sequences are generally rather similar in the different solvents,
particularly in the siliceous and boraceous cases. In general,
higher oxidation states appear to be more stable in the Na$_2$O + H$_2$O
solutions than in the others which, as shown below, implies that
H$_2$O is the weakest of the three Lewis acids considered here. The
differences would be even more marked if the aqueous solutions
were as rich in Na$_2$O as the siliceous and boraceous solutions.

In seeking to understand these differences it should be re-
membered that (1) all that distinguishes the potential of a given
ionic species in one solvent from that in another is the solvation
free energy. From this it follows that (2) the ε value of any
redox couple relative to that for the same electron transfer
process occurring in vacuum is determined by the difference in
solvation free energies for the two states. Therefore, (3) differences between ε values for a given redox couple in different
solvents reflect differences in the relative solvation energies
for the two members of the couple. Most generally these differences seem to be dependent on how effectively the higher
oxidation state member of the couple competes with the solvent

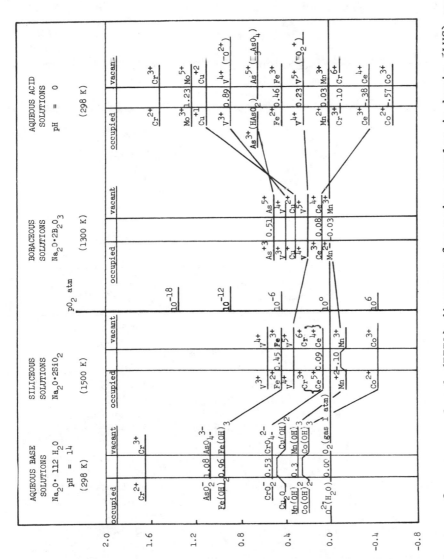

Figure 3. Electron free energy level (EFEL) diagrams for redox couples in basic (LHS) and acid (RHS) aqueous solutions compared with levels for the same couples in siliceous ($Na_2O \cdot 2SiO_2$) and boraceous ($Na_2O \cdot 2B_2O_3$) solutions at the respective temperatures of equilibration. Reference level for all cases is the oxide/oxygen level at the same temperature.

Lewis acid cation (H^+, B^{3+}, Si^{4+}, P^{5+}, etc.) for the solvent oxygens. The more effective the competition the greater the relative
stabilization of the higher oxidation state in the solvent and the
higher the ε value for the couple. Thus high oxidation states are
evidently best sought in aqueous systems (e.g. molten hydroxides)
while low oxidation states will occur most generally in high
P_2O_5 content solutions or their equivalent. The possibility, in
the case of H_2O-containing solutions, of reducing the oxide ion
activity seven orders of magnitude below that in the pure Lewis
acid by the device of adding $\sim$1 mole per liter of H^+ ions (and
thereby making a weak Lewis acid system appear relatively strong)
clearly makes the aqueous medium a very versatile one for redox
studies.

From the EFEL diagram, Fig. 3 (a)-(d), the proportion of element M in the oxidized and reduced states in a given solution in
equilibrium with 1 atm. of oxygen can be predicted using the
relation

$$[OX]/[RED] \;=\; \exp(\varepsilon/kT) \tag{5}$$

where k has the value 8.62×10^{-5} eV $\deg^{-1}$ if ε is in eV. Likewise
in a system in which exchange of electrons with O_2 gas is excluded,
the distribution of redox species between two couples e.g. Mn(II)/
(III) and Fe(II)/(III) present in the same solution can be deter-
mined by Eq. (5) using, in place of ε, the difference in ε values
between the individual couples.

In most cases, a redox couple in solution will not be in
equilibrium with oxygen at 1 atm. pressure. If $pO_2 = 0.21$ atm.,
as is often the case, the reference level is effectively shifted
by $\frac{1}{4}$ kT ln pO_2 (= 0.05 eV) and 0.05 must be subtracted from ε in
Eq. (5) before the redox ratio can be calculated (1). If the
oxygen pressure is deliberately lowered below 1 atm., e.g. by
bubbling a 1:1 CO/CO_2 mixture ($pO_2 = 10^{-11}$ atm. at 1500 K) through
the solution, then drastic changes in equilibrium distributions of
oxidized and reduced species will occur. A pO_2 scale has been
included in Fig. 3 to show the correction which must be made to
the 1 atm. O_2 electron transfer energy for solutions in equilibrium
with atmospheres of lower and higher oxygen pressures. Such
manipulations of the redox ratios are not often considered in
aqueous solution chemistry (since gas-solution equilibria are
usually only established very slowly) but are of great importance
in high temperature oxide solution chemistry. In practice, the
oxygen pressure is often only crudely controlled e.g. by adding
carbonaceous material to the solution, or melting in a "reducing"
atmosphere of unknown pO_2, with the result that predictions of
final redox ratios are not easily made. There is a great lack of
careful, controlled-atmosphere, experimentation in this field of
glass science.

3.   SOLVENT COMPOSITION EFFECTS, OXIDE TRANSFERS, AND THERMODYNAMIC SPECIES

So far, we have considered only a single solvent composition in each of our three acid-base systems, $Na_2O + H_2O$, $Na_2O + SiO_2$, and $Na_2O + B_2O_3$, (the aqueous acid case being distinct).   Further-more, we have for convenience assumed the activity of the oxide ion in Eq. (3) to be unity in each case (which can always be done by appropriate choice of standard state for the basic oxide).   If we now attempt to predict from Eq. (3) the effect of changing solvent on the position of the redox equilibrium we obtain contra-dictory results.   For instance, increase of $[O^{2-}]$ by increase of Na_2O content should shift the equilibrium expressed by Eq. (3) to the left.   In practice such a solvent composition change generally leads to an increase in the proportion of <u>higher</u> oxidation state. Similar observations in aqueous solutions were long ago rationalized by writing such electron transfer equilibria as

$$Cr^{3+} + 8OH^- \rightleftarrows CrO_4^{2-} + 4H_2O + 3e \tag{6}$$

i.e. by incorporating the transfer of an oxide ion from a solvent species to the oxidized cation as an integral part of the electron transfer process.   The strict equivalent of Eq. (6) in a silicate solvent would be

$$Cr^{3+} + 8 \, (O\text{-}\overset{|}{\underset{|}{Si}}\text{-})^- \rightleftarrows CrO_4^{2-} + 4 \, (\text{-}\overset{|}{\underset{|}{Si}}\text{-}O\text{-}\overset{|}{\underset{|}{Si}}\text{-}) + 3e \tag{7}$$

where the parallels

$$H - O - H \equiv \text{-}\overset{|}{\underset{|}{Si}}\text{-}O\text{-}\overset{|}{\underset{|}{Si}}\text{-}$$

$$(O - H)^- \equiv (O\text{-}\overset{|}{\underset{|}{Si}}\text{-})^-$$

have been drawn (H^+ and Si^{4+} are the respective Lewis acid cations).

In aqueous solutions the appropriate number of oxide transfers per electron transfer, n, is usually found to be about unity.   Eq. (7) implies the number n = 1.33 while the equilibrium in which the chromite ion CrO_2^- is recognized as the Cr^{+3} carrying species in basic solution, viz.,

$$CrO_2^- + 4 \, OH^- \rightleftarrows CrO_4^{2-} + 2 \, H_2O + 3e \tag{8}$$

implies the smaller number n - 0.67.   In principle, the appropriate number can be obtained experimentally from the slope of the plot of ε vs $\log a_{OH^-}$ ($\log a_{OH^-}$ = pH - 14) which is shown in Fig. 4(a).

The argument is given in Ref. 1 and involves the assumption that correctly chosen species mix ideally so that an equilibrium constant which is actually constant (and which therefore yields ε through Eq. 2) is obtained when K_{EQ} is written in terms of mole fractions. Fig. 4(a) is in fact just the so-called Pourbaix diagram (5) for chromium in aqueous solutions transposed to show (1) oxidation potentials referred to the oxide/oxygen reference instead of reduction potentials referred to the H_2/H^+ reference on the vertical axis, and (2) $\log a_{OH^-}$ instead of pH on the horizontal axis. Extraction of n from Fig. 4(a) requires knowledge of the variation of the reference level, ε for Eq. (2), with $\log a_{OH^-}$, which involves determination of the composition dependence of the OH^- activity. This is hardly a problem with aqueous solutions, and the reference slope is shown as a dashed line in Fig. 4(a). On the other hand, determination of the dependence of basic oxide activity on composition is a major problem for molten oxides where experimentation is very difficult and where, in consequence, results from different laboratories are often in disagreement.

Elsewhere (1) the data of Pearce (6) for CO_2 solubility in $Na_2O + SiO_2$ melts have been used, together with a standard state for the oxide solution defined by $\gamma Na_2O = 1.0$ at the composition $Na_2O \cdot 2SiO_2$, to construct a Pourbaix type diagram for the few redox couples which have been studied quantitatively in molten oxides. A portion displaying the data for the Cr^{3+}/Cr^{6+} equilibrium at 1673 K (obtained by Nath and Douglas by chemical analysis of the quenched Cr-doped $Na_2O + SiO_2$ glasses) (7) is reproduced in Fig. 4(b). The slope is greater than that shown for the CrO_2^-/CrO_4^{2-} equilibrium in Fig. 4(a) and corresponds to 1.06 oxides transferred per electron transfer, i.e. n = 1.06. Data plots for the Fe^{2+}/Fe^{3+} and the Co^{2+}/Co^{3+} equilibria in the same solvent system show somewhat smaller slopes (1) relative to ε (ref.) at the same temperature, yielding n = 0.8. The general findings are thus approximately consistent with the thermodynamic equilibrium

$$M^{n+} + \tfrac{1}{4} O_2 \text{ (g)} + \tfrac{1}{2} O^{2-} \rightleftharpoons [M^{(n+)} + 1 \, O]^{(n+)-1}, \qquad (9)$$

a specific example of which would be

$$Fe^{2+} + \tfrac{1}{4} O_2 \text{ (g)} + \tfrac{1}{2} O^{2-} \rightleftharpoons FeO^+ \qquad (10)$$

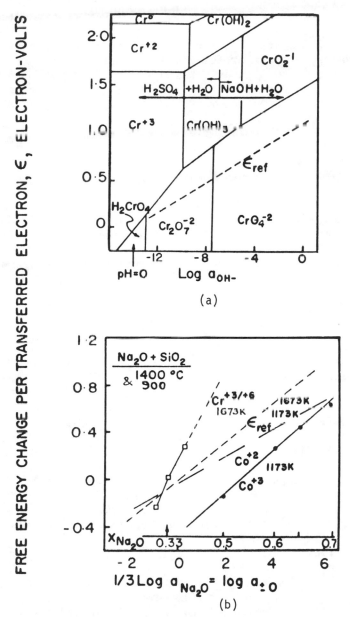

Figure 4.   (a) Free energy levels for stable chromium redox couples
(referred to oxide/oxygen reference) as a function of hydroxide activity (log scale) in aqueous solutions at 298 K.

(b)   Free energy levels for Cr^{+3}/Cr^{+6} and Co^{+2}/Co^{+3}
couples at 1673 and 1173 K respectively as a function of alkali
oxide activity (log scale) in siliceous solutions.

4.   INTERPRETATION OF OBSERVED "OXIDE TRANSFER" EQUILIBRIA

Were Eq. (10) to be taken literally, one would expect to find Fe^{3+} in an oxide glass preferentially associated with one of its near neighbor oxide ligands. While it is difficult to assess the exact coordination state of Fe^{3+} in oxide glasses because of a tendency to pass into four-fold coordination in basic solutions, the parallel case of Cr^{3+} is fairly unambiguous. Cr^{3+} can be viewed as arising through (1) an outward transfer to an electron from Cr^{2+} (which is known to exist in oxide solutions but which has only recently been characterized spectroscopically (8)) plus (2) an inward transfer of an oxide ligand. It is found, by visible and EPR spectroscopic studies (1), to be coordinated symmetrically, Cr^{3+} lying at the center of an octahedron of oxide near neighbors. There is, in other words, no preferential attachment to one oxygen. It is strongly implied that the thermodynamic species $M(III)O^+$ suggested by the redox studies are not actual species, and that Eqs. (9) and (10) require some other interpretation presumably involving electron density donation, as qualitatively argued by Duffy and Ingram (9e). We should first note that if three further electrons are transferred out from Cr^{3+} and three further "thermo-dynamic oxides" transferred in, the situation becomes better resolved, and the molecular ionic species CrO_4^{2-} can be identified spectroscopically. The spectrum of this species in the glass is, in fact, very similar to that of its well-known aqueous solution and inorganic salt counterpart. In such a case we can write with some confidence

$$Cr^{2+} \left\{ \begin{matrix} \text{ionic} \\ \text{interactions} \\ \text{with ligands} \end{matrix} \right\} + 4O^{2-} \rightarrow CrO_4^{2-} \left\{ \begin{matrix} \text{covalent} \\ \text{interaction} \\ \text{with ligands} \end{matrix} \right\} + 4e$$

in which the thermodynamic and molecular models are consistent.

It is in the intermediate situation where electron transfer has occurred but no strong covalent bonds have been generated that the problem arises. The resolution lies in the realization that wholesale transfer of electron density from molecular orbitals describing a single Si-O (or B-O, H-O) covalent bond into molecular orbitals of a metal-oxide covalent bond is unlikely for a +1 change in oxidation number unless there is some very specific molecular orbital structure favoring it (as in VO^{2+} cation formation). What is more probable is simply that the cation in its higher oxidation state finds its electrostatic competitive status, vis-a-vis the solvent Lewis acid cation, incrementally improved and will proportionately counter-polarise the electron clouds of all its near neighbor anions. The net effect will be an adjustment in electron density around the oxidized species at the expense of that around several adjacent Lewis acid cations. The ion, in

short, not only senses its electronic environment (optical basicity
(7), but also alters it as it changes oxidation number. This
incremental restoration of electron density to the oxidized species
compensates for the stepwise loss of electron density which occurs
on increase of oxidation number, thus reducing the work of transfer
of the electron away from the reduced species and, accordingly,
raising the EFE level relative to the O^{2-}/O_2(gas) reference level.

The less available the ligand electrons (i.e. the stronger
the ligand binding by the Lewis acid cation) the lower the ε value
for a couple, and accordingly the greater the preponderance of low
oxidation states for a given oxygen pressure. These conditions
are those characteristic of strongly acid media. The limit is
reached when there is no ligand electron density available at all
i.e. when the electron transfer is considered to occur in a
vacuum.* Conversely, the more polarizable the ligand the greater
the likelihood of observing higher oxidation states. Thus the
solutions in Fig. 3 in which water was the Lewis acid and H^+ the
competing cation were found to be those in which the ε levels were
highest, hence those most predisposed to dump electrons from
occupied levels (reduced species) into O_2(gas) vacant levels
thereby creating higher-valent oxidized species. In short, aqueous
NaOH solutions are the most favorable oxidation media of Fig. 3(a)-
(c) (at $pO_2 = 1$ atm) because H^+ is weak in the Lewis sense compared
to B^{3+} and Si^{4+}. The conditions most favorable of all for
generation of high oxidation numbers in oxidic systems should be
found in pure liquid oxides of large alkali cations. Such liquids
unfortunately are very corrosive and difficult to work with.
Liquid hydroxides saturated with alkali oxides probably represent
the most highly oxidizing oxidic solvent media conveniently
available.

The availability of oxide electron density to dissolved
cations thus emerges as a primary solvent characteristic in redox
chemistry. A related statement applies to the coordination
chemistry of dissolved chemical species in a single and unchanging
oxidation state (e.g. Co^{2+}, Ni^{2+}). It is shown in Ref. 1, for
instance, that the same factors which promote high oxidation numbers in redox couples promote low coordination numbers for cations
of variable coordination number.

---

*In practice such conditions are almost met in the so-called
"magic acid" solutions in which the ligands are both intrinsically
unpolarizable and also very strongly bound (as in SbF_6^-). We note
that basic data exist from which this limiting, hence reference,
EFEL diagram can be constructed.

The availability of ligand electrons can now be characterized conveniently and rather directly by optical methods following the work of Duffy and Ingram (9), and Reisfeld and Boehm (10). These workers have shown that the energy of the $^1S_o \rightarrow {}^3P_1$ transition of post-transition $d^{10}s^2$ metal ions such as Tl^+, Pb^{2+}, and Bi^{3+}, can be used as a sensitive probe of electron-donating properties of a solvent, i.e. of its basicity, and have shown that the observed transition energies can be used to arrange solvent ligands into the same nephelauxitic series obtained originally from 3d ion spectral studies. Duffy and Ingram in particular have shown that these observations can be used to define an "optical basicity" scale by which a basicity value between 0 and 1 can be assigned to each solvent. The numbers obtained can be interpreted on the basis of theoretically derived microscopic basicities assigned to each ligand according to the species to which it is coordinated. The extent to which predictions of stable melt species and their spectroscopic properties can be made using these ideas remains to be fully documented but clear parallels with thermodynamic find-ings have already emerged (9e). The optical approach has the advantage of specifying rather directly where the electrons should be, and therefore what the spectroscopic consequences of changes in solvent system, or of changes in composition within a given solvent system, should be.

The development reviewed above are promising, but much spec-troscopic and thermodynamic investigation remains to be performed before the body of ideas and experimental observation which con-stitute the "oxidic solutions" subfield of solution chemistry can be properly consolidated.

5.   SUMMARY

Similarities in the redox behavior of ions in oxide glasses (e.g. Na_2O-SiO_2, Na_2O-B_2O_3) and basic aqueous (Na_2O-H_2O) solutions expected from chemical acid-base arguments have been put on a quantitative basis by adoption of the

$$\text{oxide } (O^{2-}) \rightleftarrows \text{oxygen (gas 1 atm)} + 2e$$

electron transfer process as a reference process of zero potential for all oxidic solutions. Data comparisons are made using "electron free energy level" (EFEL) diagrams. Differences in the free energy levels for redox couples dissolved in different types of oxidic solutions (aqueous, siliceous, etc.) obtained on this basis are related to the strength of the Lewis acid (H_2O, SiO_2, B_2O_3) of the solvent system. The composition dependence of the EFEL for a given couple in each solution cannot be understood in terms of electron transfers alone. The thermodynamic data demand that "transfer in" of oxide ions from solvent to the oxidized species accompany the "transfer out" of the election(s), approximately on a one for-one basis. However, the "thermodynamic" species, such as CrO^+, FeO^+, etc. implied by such arguments, may not exist as such, since the same composition effect on the EFEL's would follow from the "transfer in" of some oxide ligand electron density from each of a number of ligands. The latter process is broadly consistent with the picture being developed by Duffy and Ingram from optical basicity arguments.

6.   ACKNOWLEDGMENT

We are grateful for the support of The National Science Foundation under Grant No. DMR73-02632A01.

7.   REFERENCES

(1)   J. Wong and C. A. Angell, "Glass: Structure by Spectroscopy," Marcel Dekker, New York, 1976, p. 241-281.

(2)   R. W. Gurney, "Ionic Processes in Solution," Dover Publ. Inc., New York, 1962.

(3a) A. Paul and R. W. Douglas, Phys. Chem. Glasses, 6, 207, 212 (1965).

(3b) W. Bancroft and R. L. Nugent, J. Phys. Chem., 33, 481 (1929).

(3c) A. Paul and D. Lahiri, J. Am. Ceram. Soc., 49, 565 (1966).

(4)   W. M. Latimer, "The Oxidation States of the Elements and Their Potentials in Aqueous Solutions," 2nd Ed., Prentice Hall, Inc., New York, 1952.

(5)   M. Pourbaix, "Atlas D'Equilibres Electrochemical Equilibria in Aqueous Solutions," Pergamon Press, Oxford, 1966.

(6)   M. L. Pearce, J. Am. Ceram. Soc., 47, 342 (1964); 48, 175 (1965).

(7)   P. Nath and R. W. Douglas, Phys. Chem. Glasses, 6, 197 (1965).

(8)   A. Paul, Phys. Chem. Glasses, 15, 91 1974).

(9a) J. A. Duffy and M. D. Ingram, J. Am. Chem. Soc., 93, 6448 (1971).

(9b)   J. A. Duffy and M. D. Ingram, J. Chem. Phys., 54, 443 (1971).

(9c)   J. A. Duffy and M. D. Ingram, J. Inorg. Nucl. Chem., 36, 39 (1974).

(9d)   J. A. Duffy and M. D. Ingram, J. Inorg. Nucl. Chem., 36, 42 (1974).

(9e)   J. A. Duffy and M. D. Ingram, J. Non-Cryst. Solids, 21, 373 (1976).

(10)   R. Reisfeld and L. Boehm, J. Non-Cryst. Solids, 17, 209 (1975).

RECENT ELECTROANALYTICAL STUDIES IN MOLTEN FLUORIDES

D. L. Manning* and Gleb Mamantov**

*Analytical Chemistry Division, Oak Ridge National
Laboratory†
Oak Ridge, Tennessee   37830

**Department of Chemistry, The University of Tennessee
Knoxville, Tennessee   37916

1.   INTRODUCTION

We have been interested for some time (1-32) in the application of modern electroanalytical methods to the study of molten
fluoride salt systems of interest to nuclear reactor technology.
The development of in situ (in-line) monitoring techniques has been
one of the main goals of this research. An obvious advantage of
in-line monitoring is that it would provide immediate knowledge of
the behavior of the reactor fuel and at the same time eliminate
sampling and other time consuming procedures. Experience with this
and other programs has generally demonstrated the value of in-line
analysis, both in more economical analyses and, frequently, in more
timely and meaningful results.

Electroanalytical methods appear to be especially attractive
for the direct analysis of electroactive species in molten salt
reactor fuel and coolant salt systems. This chapter summarizes
our electroanalytical studies of bismuth, iron, tellurium, oxide
and U(IV)/U(III) ratio determinations in molten $LiF-BeF_2-ThF_4$
(72-16-12 mole %) and $LiF-BeF_2-ZrF_4$ (65.6-29.4-5.0 mole %).   These
salts are the Molten Salt Breeder Reactor (MSBR) and Molten Salt
Reactor (MSR) fuel solvents, respectively (33).

---

†Operated by the Union Carbide Corporation for the Energy
Research and Development Administration.

By acceptance of this article, the publisher or recipient
acknowledges the U.S. Government's right to retain a non-exclusive,
royalty-free license in and to any copyright covering the article.

2. EXPERIMENTAL

The experimental set-ups for voltammetric and chronopotentiometric studies in molten fluorides have been described previously
(7,32). The controlled potential-controlled current cyclic
voltammeter is described elsewhere (34). A Tektronix 7313 storage
oscilloscope equipped with a series C-50/C-70 Polaroid Camera, and
a Hewlett-Packard Model 7045-A X-Y recorder were used to record
the voltammograms and chronopotentiograms. The melts were contained in glassy carbon crucibles (obtained from Beckwith Corp.,
Van Nuys, California). Gold, iridium, pyrolytic graphite and
glassy carbon unsheathed working electrodes were used (typical area
approx. 0.25 cm^2), as well as iridium quasi-reference electrodes
(6). The Ni(II)/Ni reference electrode (11,20,23) was not employed
in order to simulate in-line monitoring conditions involving highly
radioactive melts under which the use of this electrode may prove
impractical. The glassy carbon crucible was used as the counter
electrode. The solvent salts, $LiF-BeF_2-ThF_4$ and $LiF-BeF_2-ZrF_4$,
were prepared and purified by Chemical Technology Division personnel at the Oak Ridge National Laboratory. The procedures have
been reported previously (35). Bismuth, nickel and iron were
added as anhydrous BiF_3, NiF_2 and FeF_2, respectively. Oxide was
added as either sublimed Al_2O_3 (36) or "low fired" BeO (37). The
peroxide and superoxide were added as high purity Na_2O_2 and NaO_2,
respectively.

Tellurium compounds, Li_2Te and $LiTe_3$, were prepared by members
of the Chemistry Division, Oak Ridge National Laboratory (38).

3. RESULTS AND DISCUSSION

Bismuth

Bismuth, because of its potential use in MSBR fuel reprocessing (39), is a potential trace level impurity in a reactor fuel
stream. Under such conditions bismuth will probably be present in
the metallic state, so that some oxidative pretreatment would be
necessary before carrying out a voltammetric determination of
bismuth. The electroreduction of Bi(III) in molten $LiF-BeF_2-ZrF_4$
by voltammetry and chronopotentiometry was first studied by Hammond
and Manning (26). This work pointed to the need for additional
investigations on the behavior of bismuth mainly in the areas of
instability of Bi(III) in the melts, the extent of interference
from nickel and the feasibility of extending the limits of detection of bismuth below that of linear scan voltammetry.

As noted previously (26) stable solutions of Bi(III) in molten
$LiF-BeF_2-ZrF_4$ could not be maintained for extended periods of time

(days) in either graphite or copper cells. The two main routes for the loss appeared to be reduction of Bi(III) by the container material and/or volatilization as BiF_3. For further investigations of bismuth in molten $LiF-BeF_2-ZrF_4$, a melt containing Bi(III) at a concentration of about 10 mM was set up where the melt was contained in a glassy carbon crucible. Well defined voltammograms and chrono-potentiograms were obtained at gold, iridium and glassy carbon electrodes. The peak potential for Bi(III) reduction occurs at about -0.1 V vs Ir QRF (prior results (26) showed that this reduc-tion occurs at approx. + 0.1 V vs Ni(II)/Ni reference electrode). Again, however, the bismuth was slowly lost from the melt as revealed by a gradual decrease in the voltammetric peak current with time. To check for volatilization, a cold finger was placed in the cell for a few days; an X-ray fluorescence analysis of the deposited film revealed that the major constituent was indeed bis-muth. Thus, it now appears that bismuth is slowly lost from molten $LiF-BeF_2-ZrF_4$ primarily by volatilization; this result is in agreement with the work of Cubicciotti (40).

Because nickel is an anticipated interference, the effect of nickel on the bismuth voltammograms was determined while sufficient bismuth (approximately 5 mM) remained to produce well-defined curves. Nickel as NiF_2 was added to give a Ni(II) concentration of 15 mM. The reduction of Bi(III) precedes that of Ni(II) by ∿ 200mV under these conditions (see Figure 1, upper curve); this separation appeared to be sufficient for the determination of low concentrations of Bi(III) in the presence of typical concentrations of Ni(II) (41), particularly if derivative methods are employed.

The Ni(II) reduction wave (E_p = -.25V in Figure 1, upper curve) is poorly defined; however, two clearly separated stripping peaks were obtained on reverse scans. The first peak probably corresponds to the oxidation of Ni or possibly Ni-Bi alloy; the second peak at ∿ OV corresponds to stripping of bismuth from the electrode. When the scan is started to an initial potential of about +0.4 V vs Ir QRE, a small prewave is seen, which, although not completely understood, is believed to result from the adsorp-tion of bismuth at the electrode surface. The middle voltammogram of Figure 1 corresponds to about 10 ppm (0.11mM) Bi(III) and 240 ppm (15mM) Ni(II). The bismuth reduction wave has practically disappeared, and only one broad stripping peak (largely due to nickel) is present. The prewave, which is shown on an expanded scale in the lower curve of Figure 1, did not change significantly.

The slow loss of bismuth from the melt at 600°C was followed voltammetrically; the concentration of Bi(III) decreased from approximately 400 ppm (4.4mM) to about 7 ppm (0.077mM) in 40 days. This is about the lowest concentration that can be measured by direct linear-scan voltammetry in these melts. Anodic stripping

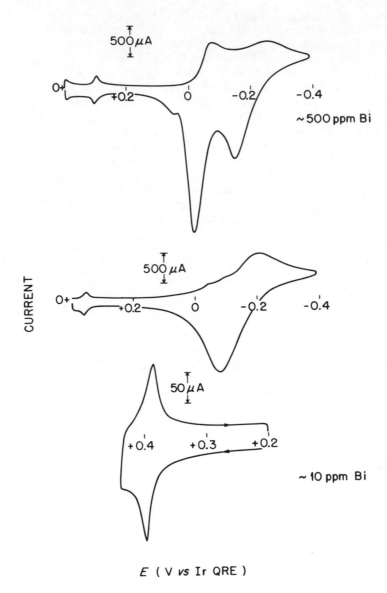

Figure 1.   Voltammograms for the reduction of bismuth(III) and nickel(II) at an iridium electrode at 500°C

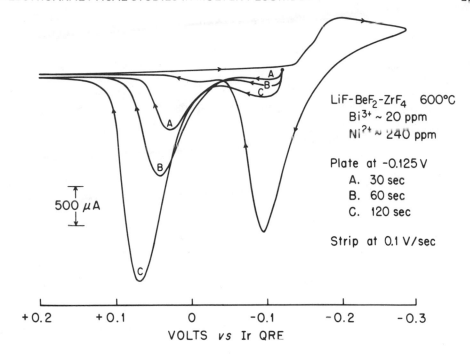

Figure 2.   Bismuth stripping curves from a glassy carbon
electrode

techniques were used for measurements of lower concentrations of
Bi(III) (41).   By plating bismuth under controlled conditions of
time and potential onto a glassy carbon electrode at a potential
sufficiently cathodic to reduce Bi(III) but not Ni(II), and then
scanning the potential anodically, the bismuth is stripped from
the electrode; the peak height of the anodic stripping curve is a
function of the concentration of bismuth.   Calibration of the
anodic stripping method was achieved by comparing the peak height
of the stripping curves to the concentration of bismuth calculated
from voltammetry and using linear extrapolation to lower bismuth
concentrations.   The peak height of the stripping curves is a
linear function of plating time (Figure 2).   The concentration of
bismuth during continued loss from the melt was followed with this
technique to sub-ppm (< 25 ppb) concentrations by employing plating
times of about 30 min.   Longer plating times were not practical
due to small signal-to-noise ratios.   Also, at longer plating
times, the interference of nickel is more severe.

The prewave observed at the iridium electrode (but not at the other working electrodes) did not change appreciably until the bismuth concentration decreased below ~ 1 ppm; then the prewave decreased markedly but not linearly with bismuth concentration. In fact, a small prewave was still observed below the detection limit of bismuth by anodic stripping. This prewave may prove of value as a qualitative indicator of bismuth.

Iron

Iron(II) is a corrosion product present in molten-salt reactor fuels. We have previously (1,4,23) carried out electrochemical studies of iron(II) in molten LiF-NaF-KF (46.5-11.5-42.0 mole %), LiF-BeF$_2$-ZrF$_4$ (69.6-25.4-5.0 mole %), and NaBF$_4$-NaF (92-8 mole %). Since the fuel solvent for the MSBR is a thorium-containing salt, LiF-BeF$_2$-ThF$_4$ (72-16-12 mole %), it was of interest to conduct voltammetric and chronopotentiometric studies of iron(II) in this fuel solvent. To determine concentration and/or diffusion coefficients by linear sweep voltammetry, it is necessary to know whether the product of the electrochemical reaction is soluble or insoluble. The measurements discussed below were done with this purpose in mind. A more detailed account of this work has been recently published (31).

A voltammogram showing the reduction of iron(II) at a gold electrode is shown in Figure 3. The circles represent the theoretical shape based on current functions tabulated by Nicholson and Shain (42) for a reversible wave where both the oxidized and reduced forms of the electroactive species are soluble. Thus, even though Fe(II) is reduced to the metal at gold, the electrode reaction very closely approximates the soluble-product case, apparently through the formation of iron-gold surface alloys. Further evidence that the Fe(II) → Fe electrode reaction at gold conforms to the soluble product case is illustrated by the chronopotentiograms in Figure 4. The ratio of the forward to reverse transition times (τ_f/τ_r) compares favorably with the predicted value of 3 (43) for the soluble case, which again points to the formation of surface alloys.

The reduction of Fe(II) at a pyrolytic graphite electrode is illustrated by the chronopotentiograms shown in Figure 5. The ratio of the transition times (τ_f/τ_r) is approximately unity (43), which is indicative that Fe(II) is reduced to metallic iron which

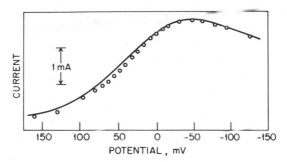

Figure 3.   Stationary electrode voltammogram for the reduction of Fe^{2+} at a gold electrode in molten $LiF\text{-}BeF_2\text{-}ThF_4$.   Potential axis is $(E - E_{1/2})/2$.   Solid line is experimental.   Circles are theoretical shape for soluble product.   Iron(II) concentration 0.027 $\underline{F}$; electrode area, 0.25 cm^2; temperature, 650°C.

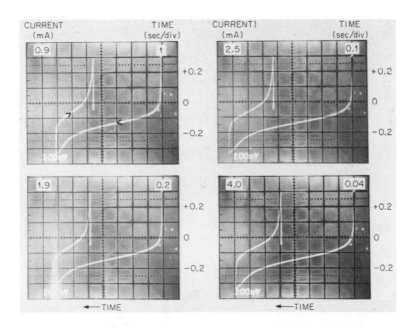

Figure 4.   Cyclic chronopotentiograms for the reduction of iron(II) at a gold electrode.   Formality of iron(II), 0.15; electrode area, 0.25 cm^2; temperature, 600°C; potential scale, volts vs Ir QRE.

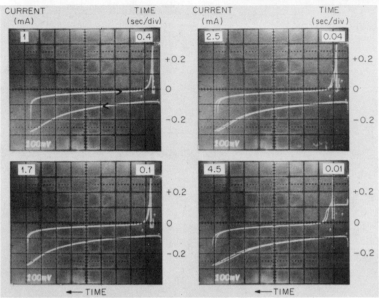

Figure 5.   Cyclic chronopotentiograms for the reduction of
iron(II) at a pyrolytic graphite electrode.   Formality of iron(II),
0.027; electrode area, 0.1 cm^2; temperature, 650°C; potential
scale, volts vs Ir QRE.

does not interact with the pyrolytic graphite and that all the iron
is stripped from the electrode upon current reversal.   The chronopotentiometric results are supported by the voltammetric difference
between peak and half-peak potentials, E_p - $E_{p/2}$ (42).   The
measured E_p - $E_{p/2}$'s are in very good agreement with the predicted
values for the reversible deposition of an insoluble substance for
n = 2 (5).   Therefore, iron appears to be reversibly reduced to a
soluble form at gold and to an insoluble material at pyrolytic
graphite.

Chronopotentiograms for the reduction of Fe(II) at an iridium
electrode at 518 and 600°C are shown in Figure 6.   The τ_f/τ_r ratio
at 518°C is approximately unity and at 600°C is 3, which is evidence that Fe(II) reduction at iridium approximates the insoluble-
species case (as with pyrolytic graphite) at 518°C and the
soluble-product case (as with gold) at 600°C.   This change in
reduction behavior with temperature was not as pronounced at gold
or at pyrolytic graphite.   Average diffusion coefficients of Fe(II)
in this melt evaluated from the chronopotentiometric measurements
by means of the Sand equation (44) are approximately 4.2 x 10^{-6},
8.0 x 10^{-6}, and 1.5 x 10^{-5} cm^2/sec at 518,600, and 700°C, respectively.

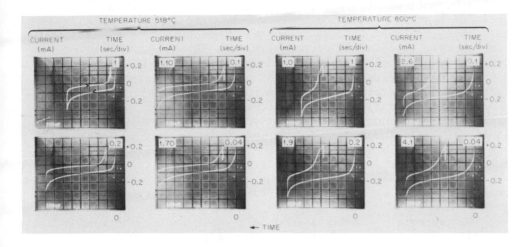

Figure 6.   Cyclic chronopotentiograms for the reduction of iron(II) at an iridium electrode.   Formality of iron(II), 0.015; electrode area, 0.2 cm^2; potential scale, volts vs Ir QRE.

Tellurium

Tellurium occurs in nuclear reactors as a fission product and results in shallow integranular cracking in structural metals and alloys (45).

Meaningful voltammograms were not obtained following additions of either Li_2Te or $LiTe_3$ ($\sim$50 mg of each in the forms of pressed pellets to $\sim$800 ml melt) to molten $LiF-BeF_2-ThF_4$ at 650°C.   It was later determined that Li_2Te is insoluble in the melt.   Chemical analysis indicated < 5 ppm (<0.13mM) Te in the melt while other solubility experiments (46) indicated < 10^{-5} mole fraction (0.3mM) of Te.   It was also found that $LiTe_3$ is not stable in the melt under our non-isothermal experimental conditions.   This substance apparently decomposes to Li_2Te and Te upon contacting the melt at $\sim$650°C.   According to the spectrophotometric measurements of Bamberger, et al. (47) isothermal conditions were necessary for the continued observation of the absorption band attributed to Te_3^- in $LiF-BeF_2-ZrF_4$; under non-isothermal conditions, the absorption band quickly disappeared with evidence of tellurium metal formation.   Behavior of tellurium in molten fluorides has also been studied by Toth (48) using absorption spectrophotometry.

In an effort to obtain additional information on the formation and stability of tellurides under non-isothermal conditions, studies were conducted on the telluride species produced in situ by cathodizing elemental tellurium ($mTe+ne \rightarrow Te_m^{n-}$).   Chronopotentio-metric and double potential step experiments conducted at a tellurium pool electrode (~ 0.1 cm^2) contained in a graphite cup revealed that the telluride species generated does not appear to be stable at $\sim 650°$C. Instability was indicated by the chronopotentiometric experiments by comparing the ratio of the forward and reverse transition times (43). Generation of a stable but insoluble substance yields $\tau_f/\tau_r = 1$; for a soluble and stable species, $\tau_f/\tau_r = 3$ is predicted. For an unstable species, on the other hand, the τ_f/τ_r ratio should be greater than three. For these experiments, the current was reversed at a time $t<\tau_f$; however, the above conclusions remain valid as long as $t<\tau_f$; however, the above conclusions remain valid as long as $t<\tau_f$. Potential-time curves recorded at the tellurium pool electrode produced $\tau_f/\tau_r>>3$ in all the runs indicating that the telluride species generated is not stable, at least within the time frame of the experiment (seconds).

In the double potential step (49) experiments, the anodic to cathodic current ratio (i_a/i_c) is plotted versus a function of time ($f(t)$) during which the potential step is applied and removed. For a stable system, $i_a/i_c = 1$ when $f(t)$ is extrapolated to zero (49). For the generation of an unstable species, $i_a/i_c<1$; this was observed for the tellurium experiments. Plots of log i versus E from the potential step experiments revealed an n value close to unity. The validity of n value determinations by this method is discussed by Armstrong, et al. (50) and by Bacarella and Griess (51). Bronstein and Posey (52) also obtained an n value of unity from polarization studies of tellurium in molten chlorides utilizing a different method.

Thus, these results indicate that an unstable telluride species is generated; under non-isothermal conditions, it undergoes a decomposition reaction. Reasonable reaction pathways are as follows:

$$2Te^{-1} \rightarrow Te_2^{2-}$$

$$Te_2^{2-} \rightarrow Te^{2-} + Te\uparrow \quad (m=1)$$

or

$$2Te_m^- \rightarrow Te^{2-} + (2m-1) \ Te\uparrow \quad (m>1)$$

The Te^{2-} ion does not appear to be soluble in fluoride melts of these compositions at least to the extent that voltammetric detection is feasible.

Recent spectrophotometric studies (48) indicate that the Te_2^{2-} ion is the most soluble tellurium species in molten fluorides but that it can only be stabilized under isothermal conditions.

Oxide and Related Species

The main purpose of these studies was to develop the basis for an in situ electroanalytical method for the determination of low levels of soluble oxide in MSR fuel streams. The importance of oxide monitoring in a molten salt reactor fuel is that oxide levels must be kept low (< few hundred ppm); otherwise, there is danger of precipitating UO_2 which could form "hot spots" in a reactor system.

Wo were also interested in establishing the electrooxidation pathway for oxide in these melts, and in this connection we also briefly studied the electrooxidation of peroxide and superoxide ions. It was observed that only at gold electrodes reproducible voltammograms for the oxidation of oxide and related species were observed. A more detailed account of this work appears elsewhere (32); the main conclusions are given below.

Cyclic voltammetric and chronopotentiometric results in molten $LiF-BeF_2-ZrF_4$ (65.6 - 29.4 - 5.0 mole%) and $LiF-BeF_2-ThF_4$ (72-16-12 mole%) in the temperature interval 500 - 710°C indicate the following electrochemical reaction pathway:

$$0^{2-} = 0 + 2e$$

$$0 + 0 = 0_2$$

$$0 + 0^{2-} = 0_2^{2-}$$

Some evidence for adsorption of 0_2 was obtained. 0_2^{2-} ions are oxidized further producing a voltammetric post-wave which increases with Na_2O_2 additions. 0_2^{2-} ions gradually decompose in these media; this decomposition is more rapid in Zr(IV)-containing melts as compared to the Th(IV)-containing melts. NaO_2 additions result in the same voltammetric results as obtained upon the addition of Na_2O_2, indicating that 0_2^- ions are more unstable in these melts than in the previously studied LiF-NaF-KF eutectic (25).

U(IV)/U(III) Ratio Determinations

The corrosion of Hastelloy N and other structural materials by MSF type fuels containing UF_4, has been attributed (53) to temperature gradient mass transfer of the most active constituent of the alloy which is chromium. The reaction is represented as

$$2UF_4 + Cr \rightarrow CrF_2 + 2UF_3.$$

The extent of this reaction is controlled by the activity of chromium in the alloy, the UF_4/UF_3 ratio in the salt and the concentration of dissolved CrF_2.

These studies were carried out in collaboration with Metals and Ceramics Division and Reactor Division personnel at the Oak Ridge National Laboratory and represent the first use of controlled potential voltammetry for molten salt corrosion studies. The principle of the voltammetric U(IV)/U(III) ratio determination (15) is illustrated in Figure 7. For these measurements, the counter electrode is the molten salt loop (see below); two iridium electrodes comprise the quasi-reference and working electrodes, respectively. Voltammograms of the U(IV) → U(III) electrode reaction are recorded relative to the Ir ORE which is poised at the equilibrium potential (E_{eq}) of the melt. This potential, in turn, is governed by the U(IV)/U(III) ratio. From the difference between E_{eq} and $E_{1/2}$ which is the voltammetric equivalent of E°, the U(IV)/U(III) ratio can be calculated from the Nernstian relationship.

In-line monitoring of the U(IV)/U(III) ratio in several corrosion test loops was carried out over a period of about two years. A schematic drawing of a loop system is shown in Figure 8. The loop portion of the assembly is constructed of Hastelloy N tubing and is heated on the bottom and left vertical side. By cooling the other two sides, the molten salt can be made to flow by density differences; thus, the term thermal convection loop. The corrosion test loops ranged from thermal convection loops of modest size to large pump driven forced convection loops. They were designed to approximate flow characteristics and temperature gradients of an MSBR. Other important features of the loops were removable corrosion specimens and ports for insertion of electrodes for voltammetric measurements.

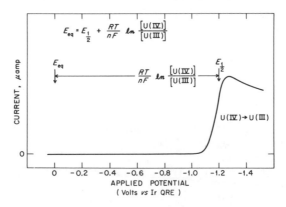

Figure 7.   Determination of U(IV)/U(III) ratio by voltammetry

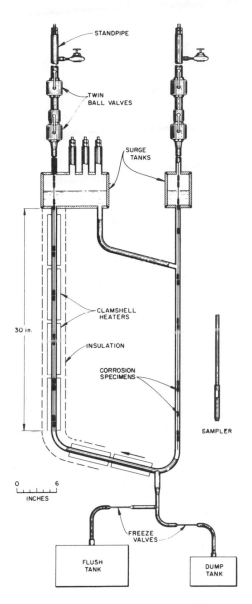

Figure 8.   Thermal convection   loop schematic*

*This Figure was originally presented at the 149th Spring Meeting of The Electrochemical Society, Inc. and is included in the publication PROCEEDINGS OF THE INTERNATIONAL SYMPOSIUM ON MOLTEN SALTS of The Electrochemical Society, Inc.

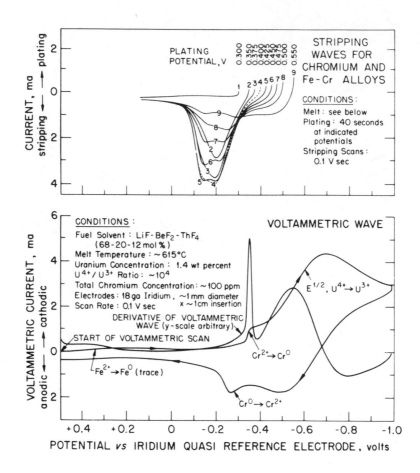

Figure 9. Typical direct and stripping voltammograms for uranium and chromium in MSBR Fuel Salt.

The use of voltammetry to provide in-line measurements of the oxidation potential and corrosion product indicators in the salt is illustrated by Figure 9, in which both the current-voltage curve and its derivative are presented. In general, well defined and reproducible voltammograms were obtained. The Cr(II) → Cr reduction wave occurs at the foot of the U(IV) reduction wave which makes precise chromium measurements difficult. Efforts to increase the sensitivity of the chromium measurements by plating and stripping techniques produced unexpected phenomena. As noted in the upper curves of Figure 9, the peak height of the stripping curve for a given plating time is critically dependent on the plating potential. It appears that plating chromium in the presence of U(IV) at potentials much more negative than the peak potential for the Cr(II) → Cr reduction wave, the concentration of U(III) at the electrode surface is sufficient to reduce part of the Cr(II) diffusing in before it is reduced by the electrode. Thus the choice of plating potentials for chromium plating and stripping experiments is relatively crucial. Nevertheless, by following relative changes in the chromium linear scan and derivative waves, useful information can be realized on the behavior of Cr(II) in these melts. Either the linear scan or derivative voltammograms can be used for $E_{1/2}$ determinations from which the U(IV)/U(III) ratio is calculated.

The variation of the U(IV)/U(III) ratio with time for two different loop materials, Hastelloy N and Inconel, is illustrated in Figures 10 and 11, respectively. For the Hastelloy N, the U(IV)/U(III) ratio reaches equilibrium at ∿100 whereas for Inconel (an alloy with a higher chromium content) the melt becomes much more reducing (U(IV)/U(III) < 10). Small perturbations usually occurred when specimens were inserted, due to an inadvertent addition of traces of moisture at the same time. These fluctuations, however, were considered to be small enough that they did not affect corrosion measurements (30).

It should be noted that in addition to the advantages of less time and expense, the in-line monitoring techniques provide information not attainable by discrete sampling methods. The in-line monitoring of the U(IV)/U(III) ratio serves as a notable example. This ratio is prohibitively sensitive to atmospheric contamination during sampling and any subsequent sample transfers to hot cells, and is rather meaningless on frozen samples because the ratio undergoes changes during cooling as a result of equilibrium shifts. We also demonstrated the feasibility of completely automating this procedure for the U(IV)/U(III) determination with a dedicated PDP-8

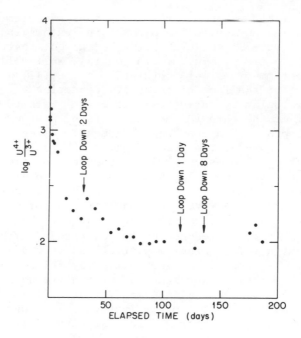

Figure 10.   Log (IV)/U(III) vs elapsed time for forced convection Hastelloy N Loop.

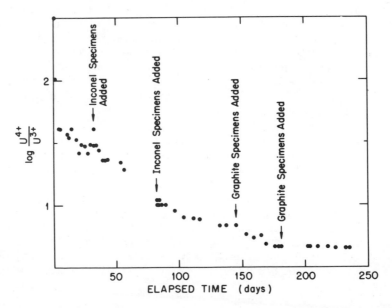

Figure 11.   Log U(IV)/U(III) vs elapsed time for thermal convection Inconel Loop.

computer (54) which operates the voltammeter, analyzes the data, and computes the U(IV)/U(III) ratio.

4. SUMMARY

Electroanalytical studies were carried out on BiF_3 in molten $LiF-BeF_2-ZrF_4$. Trace levels of bismuth are expected to occur in an MSR fuel stream from reprocessing operations. Voltammetric measurements were also made in the presence of NiF_2, an anticipated interference. Limits of detection were about 10 ppm by linear scan voltammetry but could be extended to much lower levels (< 25 ppb) by anodic stripping techniques. Stable melts containing Bi(III) cannot be maintained because bismuth is slowly lost from the melt by volatilization as BiF_3.

To determine concentration and/or diffusion coefficients by linear sweep voltammetry, it is necessary to know whether the product of the electrochemical reaction is soluble or insoluble. It was observed from voltammetry and chronopotentiometry that the Fe(II)+2e → Fe electrode reaction in molten $LiF-BeF_2-ThF_4$ closely approximates the soluble product case at a gold electrode, the insoluble product case at pyrolytic graphite, and, depending on the temperature, both soluble and insoluble product cases at an iridium electrode.

Voltammetric measurements were made in molten $LiF-BeF_2-ThF_4$ following additions of Li_2Te and $LiTe_3$ in an effort to identify soluble electroactive tellurium species. No voltammetric evidence for such compounds was obtained. Electrochemical studies were carried out on the tellurium species generated in situ in molten $LiF-BeF_2-ThF_4$ when a tellurium electrode is cathodized; the results indicated that the species generated is of the type Te_m^- (m>1) and appears to be unstable under the existing experimental conditions.

We have shown that the electrooxidation of soluble oxide species at gold electrodes in fluoride melts, such as $LiF-BeF_2-ZrF_4$ and $LiF-BeF_2-ThF_4$, provides the basis for an in situ determination of small amounts of dissolved oxide. The electrooxidation results in atomic oxygen which rapidly combines to form chemisorbed O_2 or reacts with the reactant O^{2-} forming O_2^{2-} which is oxidized further.

In-line monitoring of U(IV)/U(III) ratios was carried out for both thermal-convection and forced-convection corrosion test loops. The U(IV)/U(III) ratio reflects the redox condition of the fuel salt and stabilizes at ∼100 in loops constructed of Hastelloy N. The melt is more reducing, however, in an Inconel loop. In addition to the saving of time and expense by in-line electroanalytical techniques, information which cannot be achieved by discrete sampling methods can be obtained.

5. ACKNOWLEDGMENTS

We gratefully acknowledge assistance from and valuable dis-
cussions with the following: J. R. Keiser of the Metals and
Ceramics Division and W. R. Huntley, Reactor Division, Oak Ridge
National Laboratory who were in charge of the thermal convection
loops and forced convection corrosion test loops, respectively;
T. R. Mueller of the Instrumentation Group, Analytical Chemistry
Division, who played a very important role in the area of instrumentation design and modifications, as well as A. S. Meyer, Jr. and
J. M. Dale of the Analytical Chemistry Division who were mainly
responsible for automating the U(IV)/U(III) ratio determination
procedure.

6. POSTSCRIPT

We note with great sadness that A. S. Meyer, Jr., Group
Leader in charge of Analytical Research and Development associated
with the Molten Salt Breeder Reactor Program, passed away in
December 1975.

The Molten Salt Breeder Reactor Program at the Oak Ridge
National Laboratory was terminated in June 1976.

7. REFERENCES

(1) D. L. Manning, J. Electroanal. Chem., 6, 227 (1963).

(2) D. L. Manning and G. Mamantov, J. Electroanal. Chem., 6, 328
     (1963).

(3) D. L. Manning, J. Electroanal. Chem., 7, 302 (1964).

(4) D. L. Manning and G. Mamantov, J. Electroanal. Chem., 7, 102
     (1964).

(5) G. Mamantov, D. L. Manning, and J. M. Dale, J. Electroanal.
     Chem., 9, 253 (1965).

(6) D. L. Manning, J. M. Dale, and G. Mamantov, in Polarography,
     1964, G. J. Hills, ed., Vol. 2, p. 1143, MacMillan, London,
     1966.

(7) G. Mamantov and D. L. Manning, Anal. Chem., 38, 1494 (1966).

(8) J. P. Young, G. Mamantov, and F. L. Whiting, J. Phys. Chem., 71, 782 (1967).

(9) D. L. Manning and G. Mamantov, J. Electroanal. Chem., 17, 137 (1968).

(10) G. Mamantov and D. L. Manning, J. Electroanal. Chem., 18, 309 (1968).

(11) H. W. Jenkins, G. Mamantov, and D. L. Manning, J. Electroanal. Chem., 19, 385 (1968).

(12) H. W. Jenkins, Jr., "Electrochemical Measurements in Molten Fluorides," Ph.D. Dissertation, The University of Tennessee, March 1969.

(13) G. Mamantov, in Molten Salts; Characterization and Analysis, G. Mamantov, ed., M. Dekker, New York, N.Y. 1969.

(14) D. L. Manning and J. M. Dale, in Molten Salts: Characterization and Analysis, G. Mamantov, ed., M. Dekker, New York, N.Y. 1969.

(15) H. W. Jenkins, G. Mamantov, D. L. Manning, and J. P. Young, J. Electrochem. Soc., 116, 1712 (1969).

(16) H. W. Jenkins, G. Mamantov, and D. L. Manning, J. Electrochem. Soc., 117, 183 (1970).

(17) F. L. Whiting, "Studies of the Superoxide Ion and Other Solute Species in Molten Fluorides," Ph.D. Dissertation, The University of Tennessee, August 1970.

(18) D. L. Manning and G. Mamantov, High Temp. Sci., 3, 533 (1971).

(19) F. R. Clayton, Jr., "Electrochemical Studies in Molten Fluorides and Fluoroborates," Ph.D. Dissertation, The University of Tennessee, March 1972.

(20) H. R. Bronstein and D. L. Manning, J. Electrochem. Soc., 119, 125 (1972).

(21) F. R. Clayton, Jr., G. Mamantov, and D. L. Manning, J. Electrochem. Soc., 120, 1193 (1973).

(22) F. R. Clayton, Jr., G. Mamantov, and D. L. Manning, J. Electrochem. Soc., 120, 1199 (1973).

(23) F. R. Clayton, Jr., G. Mamantov, and D. L. Manning, High
      Temp. Sci., 5, 358 (1973).

(24) G. Ting, "Thermodynamic and Electrochemical Studies of
      Niobium in Molten Fluorides and Chloroaluminates," Ph.D.
      Dissertation, The University of Tennessee, August 1973.

(25) F. L. Whiting, G. Mamantov, and J. P. Young, J. Inorg. Nucl.
      Chem., 35, 1553 (1973).

(26) J. S. Hammond and D. L. Manning, High Temp. Sci., 5, 50
      (1973).

(27) D. L. Manning and G. Mamantov, Electrochim. Acta, 19, 177
      (1974).

(28) F. R. Clayton, Jr., G. Mamantov, and D. L. Manning, J.
      Electrochem. Soc., 121, 86 (1974).

(29) G. Mamantov, B. Gilbert, K. W. Fung, R. Marassi, P. Rolland,
      G. Torsi, K. A. Bowman, D. L. Brotherton, L. E. McCurry, and
      G. Ting, in Proceedings of the International Symposium on
      Molten Salts, J. P. Pemsler, ed., The Electrochemical
      Society, Princeton, New Jersey, 1976, pp. 234-239.

(30) J. R. Keiser, D. L. Manning, and R. E. Clausing, in Proceed-
      ings of the International Symposium on Molten Salts, J. P.
      Pemsler, ed., The Electrochemical Society, Inc., Princeton,
      New Jersey, 1976, pp. 315-323.

(31) D. L. Manning and G. Mamantov, High Temp. Sci., 8, 219
      (1976).

(32) D. L. Manning and G. Mamantov, J. Electrochem. Soc., 124, 480
      (1977).

(33) M. W. Rosenthal, P. N. Haubenreich, and R. B. Briggs, Oak
      Ridge National Laboratory Report, ORNL-4812 (1972).

(34) T. R. Mueller and H. C. Jones, Chem. Instrum., 2, 65 (1969).

(35) J. H. Shaffer, et al., Oak Ridge National Laboratory Report,
      ORNL-3951 (1964).

(36) B. Gilbert, G. Mamantov, and G. M. Begun, Inorg. Nucl. Chem.
      Letters, 10, 1123 (1974).

(37) C. E. Bamberger, H. F. McDuffie and C. F. Baes, Jr., Nucl.
      Sci. and Eng., 22, 14 (1965).

(38) D. Y. Valentine and A. D. Kelmers, Oak Ridge National Laboratory Report, ORNL-5078, 29 (1975).

(39) L. E. McNeese in ref. 33, p. 331.

(40) D. Cubicciotti, J. Electrochem. Soc., 115, 1138 (1968).

(41) A. S. Meyer, et al., Oak Ridge National Laboratory Report, ORNL-5047, 56 (1975).

(42) R. S. Nicholson and I. Shain, Anal. Chem., 36, 706 (1964).

(43) W. H. Reinmuth, Anal. Chem., 32, 1514 (1960).

(44) P. Delahay, New Instrumental Methods in Electrochemistry, Interscience, New York, 1954, p. 184.

(45) H. E. McCoy, in ref. 33, p. 207.

(46) D. Y. Valentine and A. D. Kelmers, Oak Ridge National Laboratory Report, ORNL-5132, 24 (1976).

(47) C. E. Bamberger, J. P. Young, and R. G. Ross, J. Inorg. Nucl. Chem., 36, 1158 (1974).

(48) L. M. Toth, Abstract ANAL-54, American Chemical Society Meeting, San Francisco, Calif., August 29-September 3, 1976.

(49) R. W. Murray, in Physical Methods of Chemistry, Part IIA: Electrochemical Methods, A. Weissberger and B. W. Rossiter, eds., Wiley-Interscience, 1971, p. 601.

(50) R. D. Armstrong, T. Dickinson, and K. Taylor, J. Electroanal. Chem., 64, 155 (1975).

(51) A. L. Bacarella and J. C. Griess, Jr., J. Electrochem. Soc., 120, 459 (1973).

(52) H. R. Bronstein and F. A. Posey, Oak Ridge National Laboratory Report, ORNL-5132, 29 (1976).

(53) W. R. Grimes, in reference 33, p. 111.

(54) A. S. Meyer, et al., Oak Ridge National Laboratory Report, ORNL-4449, 157 (1969).

VOLTAMMETRY OF DISSOLVED WATER AND THE ROLE OF THE HYDRATED ELECTRON IN FUSED NITRATES

Aida Espinola* and Joseph Jordan

Department of Chemistry, The Pennsylvania
State University, 152 Davey Laboratory,
University Park, Pennsylvania 16802

1.   INTRODUCTION

In the early 'sixties, Swofford and Laitinen [1,2] first reported that small amounts of water dissolved in molten alkali nitrate solvents yielded steady-state current-voltage curves whose
sigmoid shape resembled classical polarograms which exhibited a well-
defined limiting current domain. This phenomenon has since been
referred to as the so-called "water wave" [3] and reinvestigated by
several authors at mercury, platinum and gold indicator electrodes
[4-6]. Using Levich's rotated platinum disk electrode (RDE), Peleg
[5] has successfully made use of the "water wave" for the quantitative determination of moisture in nitrate melts. In a preliminary investigation, T.E. Geckle in our laboratories [7] has sub-
stantiated that the relevant limiting currents at the RDE were indeed proportional to the partial pressure of water in the supernate. However, the same investigation revealed that neither hydrogen nor oxygen were produced during the electrolysis. Results of a
comprehensive study of the water wave phenomenon are presented and
discussed in this paper, including voltammetry at the RDE and controlled potential coulometry, complemented by analysis of reaction
products via appropriate chemical methods and mass spectrometry.
A mechanism is postulated, invoking involvement of a hydrated electron species, which accounts for the paradoxical experimental findings previously reported in the literature.

---

*Based on a Ph.D. thesis by Aida Espinola (present address: Univ.
Fed. Rio de Janeiro, Coppe-Mt, C Postal 1191 ZC-00, 20.000 -
Rio Janeiro, Brasil).

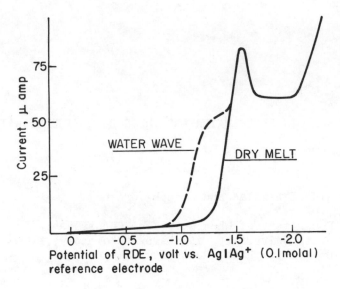

Figure 1. Typical Current-Voltage Curves

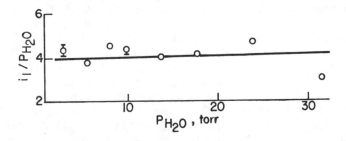

Figure 2. Verification of Proportionality Between
Limiting Currents (i_l) and Partial Pressure of Water (p_{H_2O})

2.   EXPERIMENTAL

Several hundred cathodic current voltage-curves were recorded at 250°C (=523°K) in an equimolar $NaNO_3$-KNO_3 melt. Water concentrations ranged from 0.001 molal to 0.013 molal, corresponding to partial pressures of water between 2 and 30 torr in the supernate. A controlled potential three-electrode polarograph was constructed ad hoc and equipped with enhanced current measuring capabilities, incorporating computer electronics [8,9]. The rotating platinum disk indicator electrode, the auxiliary platinum counter electrode (which carried the current), the Ag/Ag$^+$ reference electrode, the electrolysis cell and ancillary equipment (e.g., furnaces, ovens and thermoregulators used to maintain the temperature at 523° ± 0.5°K) have been described in previous papers [10-12].

For coulometry, potentials were controlled with a Wenking Model 61-TR Potentiostat. Current-time integrals were recorded as the output of conventional analog feedback devices, by accumulating the feedback current on a capacitor with the aid of an appropriate operational amplifier circuit [3].

Absence of hydrogen and oxygen evolution during electrolysis was ascertained with an A.E.I.-MS902 mass spectrometer. Other monitored electrolysis products included nitrite and hydroxide. NO_2^- was determined by UV spectrophotometry, using an incremental addition method developed by one of us (AE) and described elsewhere [13]. Hydroxide was titrated acidimetrically to a conventional indicator end point. Both nitrite and hydroxide were determined after quenching the melts and dissolving appropriate aliquots in water.

3.   RESULTS

A typical "water wave" is illustrated in Figure 1. The crucial qualitative finding was that no H_2 or O_2 were electrogenerated in the potential range of the limiting current. In this context, mass spectrometric findings were conclusive at the 99% confidence level. Painstakingly careful analysis of products and of coulometric current-time integrals revealed that the overall electrode reaction was:

$$NO_3^- + H_2O + 2e = NO_2^- + 2OH^- \tag{1}$$

On the other hand, the limiting currents were strictly proportional to the concentration of dissolved water. The linear dependence between limiting currents (i_l) and partial pressure of water (p_{H_2O}) is documented in Figure 2. The solubility equilibrium of water between the melt phase and the water phase obeyed Henry's Law. Henry's Law Constant was 5.4×10^{-4} mole H_2O per kg per torr

Table I

Diffusion Coefficients Calculated from Equation 2,[*]
Substituting Known Water Concentrations for C

$\underline{C}$ (moles per kg)	$\underline{D}$ at 523°K (cm^2/sec)
$1.28 \cdot 10^{-3}$	$(5.4 \pm 0.3) \times 10^{-5}$
$4.74 \cdot 10^{-4}$	$(5.5 \pm 0.1) \times 10^{-5}$

[*]$A = 1.38 \cdot 10^{-3} \ cm^2$; RDE rotated at 900 RPM

at 250°C under the prevailing experimental conditions. Likewise, in a range between 400 and 900 RPM, limiting currents were proportional to the square root of the rate of rotation of the indicator electrode in accordance with the Levich Equation:

$$i_1 = 0.62 \ nFA \ \omega^{1/2} \ \nu^{-1/6} \ D^{2/3} \ C \tag{2}$$

where $\underline{n}F$ denotes the number of coulombs per mole, $\underline{A}$ and $\underline{\omega}$ the area and angular velocity of the RDE, $\underline{\nu}$ the kinematic viscosity of the solution; and $\underline{D}$ and $\underline{C}$, respectively, are the diffusion coefficient and bulk concentration of the electroreactive species. Substituting in Equation 2 known concentrations of water, the diffusion coefficient assignments listed in Table I were evaluated from measured limiting currents.

The water wave was found to shift along the potential axis as a function of nitrite and hydroxide concentrations added ab initio to the melt.* The observed potential shifts are accounted for by the fact, that in the presence of appreciable bulk concentrations of nitrite and hydroxide the "wave equation" of Reaction 1 has the form:

$$E = E' + \frac{RT}{2F} \ln \frac{i_1 - i}{(a+i)(b+i)^2} \tag{3}$$

---

*In this context, it should be noted that all the nitrate melts, in which we observed water waves, contained trace amounts of nitrite as a contamination, probably due to unavoidable thermal decomposition of nitrate.

if the process $\underline{is}$ Nernst-reversible, and

$$E = E' + \frac{RT}{2\alpha F} \ln \frac{i_1 - i}{(a+i)(b+i)^2} \qquad (4)$$

if it is $\underline{not}$ Nernst-reversible. In Equations 3 and 4, the quantities $\underline{a}$ and $\underline{b}$ denote "current equivalents" of the bulk concentrations of nitrite and hydroxide present in the melt. Plots based on experimental curves of $\underline{E}$ versus the quantity $\log [(i_1 - i)/(a+1)(b+1)^2]$ yielded a slope of (64 ± 1) millivolt. This indicates unambiguously that the electrode reaction was $\underline{not}$ $\underline{Nernst-reversible}$, because

$$\frac{2.3\ RT}{2F} = 51.7 \text{ millivolt at the prevailing temperature of } 523°K \qquad (5)$$

Consequently, Equation 4 holds, yielding:

$$\frac{2.3\ RT}{2\alpha F} = (64 \pm 1) \text{ millivolt at } 523°K; \\ \alpha = 0.81 \pm 0.02 \qquad (6)$$

In view of this finding, viz., that Reaction 1 was $\underline{irreversible}$, the electrochemical rate constant, $\underline{k}$, was evaluated at various potentials from an applicable relationship [14]:

$$k = \frac{i}{i_1 - i} 0.62\ \nu^{-1/6}\ \omega^{1/2}\ D^{2/3} \qquad (7)$$

Electrochemical rate theory predicts, quite generally, the following correlation between k-s and potentials:

$$k_E = k_o \cdot e^{-\alpha E\ nF/RT} \qquad (8)$$

where the subscript zero denotes an arbitrary reference potential. Based on Equation 8, a plot of $\underline{\log k}$ versus $\underline{E}$ should be rectilinear with a slope consistent with Equations 4 and 6. Equation 8 was indeed verified in all our relevant experiments. A comparison of the transfer coefficient assignments, based on Equation 4 on the one hand, and on Equation 8 on the other hand, is presented in Table II. The agreement appears reasonable.

Table II

Comparison of Transfer Coefficients

Calculations Based on	Value of α
Equation 4	0.81 ± 0.02
Equation 8	0.85 ± 0.02

4. DISCUSSION

Our experimental findings are generally consistent with what has previously been reported in the literature. In particular, we agree with a recent significant observation by Lovering et al. [6], that the presence of some nitrite in the melt appears to be a necessary condition for observing the "water wave." It seems that there is little controversy about the experimental facts. As far as the formulation of a plausible mechanism is concerned, we feel that it should be consistent with the following experimental results:

(a) The overall electrode reaction must have the stoichiometry explicited in Equation 1;

(b) The proportionality between limiting currents and water concentrations must be accounted for.

We suggest that an observation by Swofford [1], viz., that a blue color appeared transiently in the vicinity of the indicator electrode, is a clue to the nature of the intermediate which may be implicated. We have also observed that blue color. Accordingly, we postulate that the blue color was due to the formation of a hydrated electron species by an <u>electrode reaction</u> of the type:

$$H_2O + e^- = [e \cdot H_2O]^- \qquad (9)$$

In turn, the aquated electron was the reducing agent involved in the conversion of nitrate to nitrite by the <u>chemical reaction</u>:

$$NO_3^- + 2[e \cdot H_2O]^- = NO_2^- + 2OH^- + H_2O \qquad (10)$$

Reaction 1 represents the obvious summation of Reaction 9 (taken twice) plus Reaction 10. Reaction 9 is assumed to be the rate determining step, controlled--in turn--by the mass transport of water from the bulk of the melt to the electrode interface. This accounts for the proportionality of limiting currents to water concentration. Indeed, the diffusion coefficient assignments listed in Table I are reasonable values for water dissolved in the melt. The relatively high transfer coefficients in Table II indicate appreciable similarity between the product of Reaction 9 (the hydrated electron) and an activated complex (transition state) which is its immediate precursor. We concur with Lovering [6] that spectroscopic studies are necessary to substantiate that the blue color was indeed due to hydrated electrons. To account for the observation that ab initio presence of traces of nitrite is a necessary prerequisite of the water wave, we speculate that Reaction 10 occurred via the following sequence:

$$NO_2^- + [e \cdot H_2O]^- = NO_2^{2-} + H_2O \tag{11}$$

$$NO_2^{2-} + NO_2^- = N_2O_4^{3-} \tag{12}$$

$$N_2O_4^{3-} + H_2O = N_2O_3^- + 2OH^- \tag{13}$$

$$N_2O_3^- = NO_2^- + NO \tag{14}$$

$$NO + [e \cdot H_2O]^- = NO^- + H_2O \tag{15}$$

$$NO^- + NO_3^- = 2NO_2^- \tag{16}$$

5.   REFERENCES

(1)   H.S. Swofford, Jr., Ph.D., thesis, University of Illinois, 1962.

(2)   H.S. Swofford, Jr., and H.A. Laitinen, J. Electrochem. Soc., 110, 814 (1964).

(3)   J. Jordan, J. Electroanal. Chem., 29, 127 (1971).

(4)   P.G. Zambonin, J. Electroanal. Chem., 24, 365 (1970); ibid., 24, App. 25 (1970).

(5)   M. Peleg, J. Phys. Chem., 71, 4553 (1971).

(6)   D.G. Lovering, R.M. Oblath, and A.K. Turner, J. Chem. Soc.
      Chem. Comm., 1976, 673.

(7)   T.E. Geckle, M.S. thesis, Pennsylvania State University,
      1964.

(8)   D.J. Fisher, W.L. Belen, and M.T. Kelley, Chem. Instrum.
      1, 211 (1968).

(9)   J. Jordan, USAEC Rept. T10-20114 (1964).

(10) P.G. Zambonin and J. Jordan, Anal. Letters, 1, 1 (1967).

(11) P.G. Zambonin and J. Jordan, J. Amer. Chem. Soc., 89, 6365
      (1967); ibid., 91, 2225 (1969).

(12) P.G. Zambonin, Anal. Chem., 41, 868 (1969).

(13) A. Espinola, Anal. Letters, 8, 627 (1975).

(14) V.G. Levich, Physicochemical Hydrodynamics, Prentice-Hall,
      Englewood Cliffs, N.J., 1962.

SUBJECT INDEX